WASTEWATER COLLECTION SYSTEMS MANAGEMENT

WEF Manual of Practice No. 7
Fifth Edition

Prepared by **Task Force on Wastewater Collection Systems Management**
Stephen Tilson, *Chair*

Danny Akagi
Rosanne Cardozo
Dana Carroll
Stephen E. Cordes
Lewis Debevec, Jr.
Wayne Dillard
Robert Domkowski
Joseph C. Filippone, Jr.
David A. Flowers
John Gresh
James Heaney
Earl Jenkins
Ralph G. Johanson
James Kent
Jim McGregor
Jim L. Meara
Greg Miller
Tom M. Owens
Douglas Prentiss
Reggie L. Rowe
Paul Schuler
Yu-Chun Su
Paul Vogel
Noah Walker
Christopher D. Wilson

Under the Direction of the **Municipal Subcommittee of the Technical Practice Committee**

1999

Water Environment Federation
601 Wythe Street
Alexandria, VA 22314–1994 USA

IMPORTANT NOTICE

The material presented in this publication has been prepared in accordance with generally recognized engineering principles and practices and is for general information only. This information should not be used without first securing competent advice with respect to its suitability for any general or specific application.

The contents of this publication are not intended to be a standard of the Water Environment Federation (WEF) and are not intended for use as a reference in purchase specifications, contracts, regulations, statutes, or any other legal document.

No reference made in this publication to any specific method, product, process, or service constitutes or implies an endorsement, recommendation, or warranty thereof by WEF.

WEF makes no representation or warranty of any kind, whether expressed or implied, concerning the accuracy, product, or process discussed in this publication and assumes no liability.

Anyone using this information assumes all liability arising from such use, including but not limited to infringement of any patent or patents.

Library of Congress Cataloging-in-Publication Data

Wastewater collection systems management / prepared by Task Force on Wastewater Collection Systems Management ; under the direction of the Municipal Subcommittee of the Technical Practice Committee. — 5th ed.

p. cm. — (WEF manual of practice ; no. 7)

Includes bibliographical references and index.

ISBN 1-57278-152-1

1. Sewage disposal plants—Management. 2. Sewage disposal plants—Maintenance and repair. I. Water Environment Federation. Task Force on Wastewater Collection Systems Management. II. Series: Manual of practice ; no. 7.

TD201.W337 no. 7 1999

[TD746]

628.1′68 s—dc21 99-13015

[628.3′068] CIP

ISBN 1-57278-152-1

Printed in the USA **1999**

Water Environment Federation

Founded in 1928, the Water Environment Federation is a not-for-profit technical and educational organization. Its goal is to preserve and enhance the global water environment. Federation members number more than 42 000 water quality professionals and specialists from around the world, including engineers, scientists, government officials, utility and industrial managers and operators, academics, educators and students, equipment manufacturers and distributors, and other environmental specialists.

For information on membership, publications, and conferences contact

Water Environment Federation
601 Wythe Street
Alexandria, VA 22314-1994 USA
1-703-684-2400
http:\\www.wef.org

Manuals of Practice

(As developed by the Water Environment Federation)

The Water Environment Federation (WEF) Technical Practice Committee (formerly the Committee on Sewage and Industrial Wastes Practice of the Federation of Sewage and Industrial Wastes Associations) was created by the Federation Board of Control (now Board of Directors) on October 11, 1941. The primary function of the committee is to originate and produce, through appropriate subcommittees, special publications dealing with technical aspects of the broad interests of the Federation. These manuals are intended to provide background information through a review of technical practices and detailed procedures that research and experience have shown to be functional and practical.

A.B. Pincince
T.E. Sadick

Authorized for Publication by the Board of Directors
Water Environment Federation

Quincalee Brown, *Executive Director*

Contents

List of Tables

List of Figures

Preface

Collection systems operation requires a high level of competence, training, education, and experience to perform to design specifications. Over the years, the collection systems field has grown because of public awareness of health hazards, clean water issues, and groundwater pollution. Many states now require certified workers to be in charge of collection systems operations.

This manual is a revision of the 1992 Manual of Practice (MOP), *Wastewater Collection Systems Management* (MOP 7), and represents a joint effort by the Water Environment Federation's Collection Systems and Technical Practice Committees. Under a cooperative agreement with WEF, the U.S. Environmental Protection Agency (U.S. EPA) provided grant assistance for the project. The U.S. EPA project officer for the cooperative agreement was Charles Vanderlyn. With the exception of minor editorial changes, Chapter 4, "Pumping Stations," was not updated for this edition.

This manual serves as a systems guide and includes principles of operation and management that have been developed over the years. This manual is not a "how to" book detailing the operation of various machines and techniques; it is a guideline for assisting collection systems managers in establishing or reorganizing their systems operations. Included in the manual are procedures, practices, and guidelines for effectively operating, maintaining, and establishing or reestablishing wastewater collection systems as dependable public utilities.

This manual is designed to address a broad range of operation and maintenance conditions. Topics include basic policies, procedures for routine operations, preventive maintenance, emergency service and repairs, management planning, recordkeeping, personnel requirements and training, system design, contracting, safety issues, and capital asset management.

Though many issues and topics are discussed in this manual, in-depth information may be desired for a specific project or plan. Each chapter includes a reference section on suggested reading materials on various items for more in-depth information. For additional reading pertaining to collection systems, refer to the following Federation publications: *Confined Space Entry*, *Design and Construction of Urban Stormwater Management Systems* (MOP FD-20), *Design of Wastewater and Stormwater Pumping Stations* (MOP FD-4), *Existing Sewer Evaluation and Rehabilitation* (MOP FD-6), *Gravity Sanitary Sewer Design and Construction* (MOP FD-5), and *Safety and Health in Wastewater Systems* (MOP 1).

Principal authors of the manual are

Richard W. Arbour (Chapter 4)
Stephen E. Cordes (Chapter 6)
Joseph C. Filippone, Jr. (Chapter 5)
David A. Flowers (Chapter 7)
D. Thomas Isley (Chapter 6)
Ralph G. Johanson (Chapter 8)
Douglass B. Prentiss (Chapter 1)
Reggie L. Rowe (Chapter 2)
Stephen B. Tilson (Chapter 3)

In addition to the WEF Task Force and Technical Practice Committee Control Group members, reviewers included Charles Vanderlyn.

Authors' and reviewers' efforts were supported by the following organizations:

3-Waters Technical Services, Solana Beach, California
Black & Veatch LLP, Kansas City, Missouri
Burns & McDonnell, Kansas City, Missouri
Carollo Engineers, Portland, Oregon
Carroll Engineering, Belvidere, Illinois
CH2M Hill, Oakland, California
David A. Flowers, P.E., Cedarburg, Wisconsin
Gainesville Regional Utilities, Gainesville, Florida
Greeley and Hansen, Chicago, Illinois
GRW Engineers, Inc., Louisville, Kentucky
ITT Flygt Corporation, Trumbull, Connecticut
Killam Associates, Millburn, New Jersey
Montgomery Watson, Coral Gables, Florida
Narragansett Bay Commission, Providence, Rhode Island
Northeast Ohio Regional Sewer District, Cleveland, Ohio
PBS&J (formerly Espey, Huston & Associates, Inc.), Austin, Texas
RJN Group, Norcross, Georgia
U.S. Environmental Protection Agency Headquarters, Washington, D.C.
University of Colorado at Boulder, Boulder, Colorado

Chapter 1
Safety

SAFETY PROGRAM ELEMENTS

The selection and implementation of an effective safety program is an important management responsibility. Successful wastewater collection system safety programs involve employees in the identification and elimination of unsafe working conditions. The positive results of these efforts are reduced costs and efficient operations. The successful manager of today's wastewater collection operations understands the costs associated with on-the-job injuries, damage claims, and vehicular accidents. Safety is a management issue like any other and requires professional administration.

The physical makeup of the collection system is a key factor in evaluating which type of safety program is necessary for an organization. Many large metropolitan cities have hundreds of lift stations with pipes large enough to drive a car through. A successful safety program for this operation will be much different from a small city operating gravity lines. However, there are key elements to be considered by all organizations. This chapter attempts to identify those safety considerations common to all workers in the wastewater collection industry.

A successful safety record can be achieved through the creation of a plan that consists of the following ten basic elements:

- Safety team,
- Safety suggestion systems,
- Safety training,
- Safety incentives,
- Safety recognition,
- Group safety meetings,
- Public safety programs,
- Emergency preparedness,
- Employee health programs, and
- Safety regulations.

SAFETY TEAM. In smaller organizations the safety team may comprise the manager and lead workers. These individuals meet on a regular basis to focus on safety issues. These activities are then shared with other workers on a daily basis. The team members must lead by example and demonstrate the importance of safety in the workplace. These types of safety teams may benefit from membership in local safety councils or trade associations.

Other sources of safety-related activities and information are trade publications on safety. In addition, many state safety agencies publish monthly newsletters, which can be valuable assets to smaller organizations. In some states, cooperative associations made up of smaller organizations have formed to share resources.

Just as locations, equipment, and manpower are discussed, safety must be an important part of the planning activities for every job. If employees know that working safely is an important part of their responsibilities, then they will try to do so.

An effective safety team for a large wastewater collection system should comprise representatives from all areas of the organization. Called safety committees or safety teams, these groups are the foundation of a successful safety program. Committee or team members typically represent the following wastewater collection areas:

- Heavy construction,
- Television inspection and sealing,
- Equipment operation and maintenance,
- Lift stations,
- Electrical and mechanical maintenance,
- Management,
- Union,
- Safety, and
- Drivers.

Once again, the makeup of the system and the personnel that maintain it are the keys to a successful safety team or committee. A large system may have specialized groups such as confined space "rescue teams," and a member of that team should be on the safety committee.

Rotating membership so that personnel serve alternate terms helps ensure representative employee participation. Meetings should be held regularly no matter what the group size. Larger organizations should keep written minutes of each meeting. All items can be tracked in writing until resolved or completed. Each meeting starts with a review of old business and an update of its status. After all old and current business is reviewed, each member is given the opportunity to discuss any new business they have been asked to bring forward.

Basic responsibilities of the safety committee in the case of an accident are to prevent a recurrence and to determine responsibility for the accident. This typically is accomplished by a review of the written report filed for all accidents. The safety committee reviews each of these reports to ensure that proper steps are taken to prevent a recurrence.

In some organizations, the safety committee may make recommendations as to the effect an accident will have on the individual's safety incentive (see Safety Incentive Programs). In other cases, committee members may be assigned to investigate some aspects of the report.

SAFETY SUGGESTION SYSTEMS. A safety suggestion program serves many useful purposes. It allows an employee to resolve a safety-related problem that has not been resolved through the normal chain of command. Safety suggestion programs increase the safety awareness of employees and encourage participation beyond members of the safety committee. Giving employees the opportunity to contribute to the safety program by making suggestions increases each employee's sense of responsibility for his or her safety and that of those around them. A suggestion program multiplies the resources of the safety committee for identifying and preventing potential problems by adding the "eyes and ears" of all employees.

Most importantly, written safety suggestions require a response and create a path for follow-up. Forms on which employees can make safety suggestions can be located throughout the workplace for safety committee review and response to each suggestion. In smaller organizations, the manager may simply implement the suggestion and give credit to the originator. This will demonstrate the commitment of the manager to safety and encourage the other employees to participate in safety-related activities. Larger organizations with safety committees might take some or all of the following actions in responding to suggestions:

- Evaluate the suggestions and recommend implementation,
- Assign committee members to investigate suggestions and provide a report at a future meeting,

- Provide written recommendations to management for implementation of suggestions that require budget modifications,
- Review all suggestions to determine if special recognition is warranted for the individual submitting the report,
- Present all suggestions and their dispositions at group safety meetings, and
- Prepare a written response to the suggestion originator.

A suggestion system must be supported to be effective. Many programs do not work because they are not implemented by staff members familiar with the importance of controlling the unwanted costs associated with accidents. The best way to control safety-related costs is to listen to the people who operate the system—they know which improvements need to be made and must be encouraged to provide this information.

SAFETY TRAINING. One of the most typical reasons for fatalities in wastewater collection systems is the failure of victims to recognize hazards. Safety training cuts across all job descriptions and emphasizes the need to recognize and deal with hazards. There are many programs required by regulation regardless of the size of the system. In most cases, these programs must be in writing and the associated training must be well documented. Failure to properly document training may result in citations. Proper documentation may preclude unnecessary inspections and work disruptions.

The following list identifies the basic programs and areas of concern typical to wastewater collection systems, although state and local safety officials should be consulted for more detailed guidelines:

- Written safety program;
- Confined space program;
- Lockout/tagout program;
- Training of personnel in trenching and excavations competency;
- Respiratory protection program;
- Biological hazards in wastewater;
- Traffic control and work site safety;
- Chemical handling (hazardous communication or right-to-know);
- Material safety data sheets;
- Electrical, mechanical, pneumatic, and hydraulic systems safety;
- Cardiopulmonary resuscitation (CPR) and first aid; and
- Commercial driving licenses (CDLs).

A safety training officer may be a shared resource person or a full-time employee in a wastewater collection system, depending on the size of the system. In either case, the basic responsibilities are similar. In larger organizations the safety training officer serves as the coordinator for safety activities and is available 24 hours per day to respond to emergencies and special training needs. Typical duties routinely handled by this person may include coordinating other training professionals, performing hands-on demonstrations during safety training, and performing field inspections. This person typically is the individual other than the department manager who is responsible for the implementation of policies and procedures within the workforce. Department managers and training officers who work closely together to accomplish common goals will be successful in the majority of their tasks. Where the size of the collection system will not support a full-time person in this position, it is typical for a risk-management employee to be assigned the task of incorporating the wastewater collection system into his or her area of concern. In such cases, the safety person may only handle difficult claims or accidents, and the department manager is left with the daily responsibility of the implementation of safety. The risk-management representative should still be viewed as a safety resource for the department manager and may coordinate training with other areas or notify operational managers of changes in safety regulations.

Issues such as CDL requirements should be identified and guidelines distributed within operational areas by someone in the highest administrative areas of all organizations. Issues such as drug testing for CDL drivers must be coordinated at the highest level of an organization. The main challenge for today's manager is to communicate efficiently with the upper levels of the organization to know what new things are happening, and to then know the system well enough to be able to implement those changes.

Perhaps the most needed and overlooked supervisor's training program deals with how to create safety training modules. Training should result in the attainment of a measurable performance objective. Student task demonstrations that describe performance objectives form the basis of the training module. For those cases in which knowledge is all that is required, a simple written test may be all that is needed to demonstrate employee proficiency.

There are, however, many safety training requirements that will not be satisfied by a written test. In such cases, a training module must be developed in which students are told what is expected and given the necessary equipment to perform the task. For confined space entry, necessary equipment includes a blower, gas detector, tripod and winch, and full-body harness. Where necessary, a procedure may also be included on the module outline, with separate lists for tools, resources, or other materials needed by the student.

A description of the student's task may also be provided. In the confined space entry example, the student might be told to perform a safe confined space entry using all appropriate safety precautions. The training module may also describe how well the student is expected to perform the task, which may be written, timed, or directly monitored by the course manager. In many cases, the training itself may be hazardous and a person competent in the skill being taught must instruct the class. Test results should be documented and filed after completion. Keep in mind that these training records need to be available during safety inspections.

Other departments may also use training modules. Training modules on traffic control, trenching and excavation, right-to-know information, and other topics can be shared by different departments. A well-organized training program will incorporate safety training and can save time if properly documented. Employees should be trained to perform the work in the safest manner. Training is required before employees perform any work in hazardous areas.

SAFETY INCENTIVE PROGRAMS. Incentive programs recognize employees who perform the most hazardous work. Specific measurable goals, such as decreases in lost time from accidents, auto liability, and general liability claims, are necessary for a successful incentive program.

Incentive systems must be based on positive safety performance. The amounts budgeted for safety incentive programs vary with the size of the system. However, simple formulas may be developed for budget purposes. For example, a system employing 200 workers could offer a $50-per-person incentive if $10 000 were allocated to the program.

Funding for the program correlates with the results. A good year results in an increase in the amount budgeted. If a program started with $10 000 and resulted in a 10% reduction in lost time from accidents, a reasonable amount to budget for the next year might be $11 000. A bad year might result in a reduced amount of $9 000. It is important to tie overall performance to overall rewards. Other programs issue incentives based on the actual hazardous procedures performed by the employees. Although each of these approaches has good and bad points, the soaring costs of medical bills and litigation have prompted many organizations to take a far more aggressive role in the use of safety incentives. In an unusual turn of events, the smaller organizations seem to be benefitting more than large systems from these programs. In programs where one employee has many different types of responsibilities, that employee also has many exposures to unwanted incidents. Additionally, the direct costs and lost productivity associated with accidents may weigh much heavier on systems with fewer resources. At today's costs, the prevention of just one accident can pay for a safety incentive program. It is far easier to improve system economics by reducing unwanted costs than it is to increase revenues. In addition, the reduction of unwanted incidents results in lower costs for workers compensation, damage claims, and lost production. The proper administration of an effective safety program will allow the company to retain more of its profits.

SAFETY RECOGNITION PROGRAM. It is important that employees be rewarded for attention to safety issues. This may be accomplished by a manager simply telling the entire workforce what was suggested and who suggested it, and making sure it is implemented. In some cases, special recognition in the form of a lunch, dinner, or special program serves to encourage others and rewards positive behavior. In the case of exceptional actions, an award or plaque presented publicly can help send a positive message to the workforce. In larger organizations, other methods are used to provide recognition. Safety committee members may nominate employees who should be acknowledged as having demonstrated or performed safe work practices. Safety incentive points may be awarded to the most outstanding nomination where a program is in place.

GROUP SAFETY MEETINGS. A general safety meeting should be held on a regular basis to review the activities of the safety committee. Awards may be distributed, and safety suggestions that have been submitted are reviewed so that all employees understand the efforts underway. This approach facilitates open discussion and identification of other safety-related activities. This may also be a training opportunity. Short informational general programs on topics such as proper lifting, health, or security help keep employees focused on positive issues. Some organizations have meetings every morning to schedule the daily work assignments. While this is a good practice, it should not take the place of a group safety meeting. Many organizations spend so much time promoting efficiency that work tends to take priority over safety in the minds of workers. Take the time to have regular meetings that focus just on the safety issues that affect the organization. It is time well invested, and the rewards are reduced costs and increased revenues.

PUBLIC SAFETY PROGRAMS. Again, the size of the system is a key in deciding what is appropriate for the organization. Smaller systems may not have the resources to actively promote training in the community but may be able to participate in public programs that benefit the community. Examples of these programs are crime-watch programs in which a radio-equipped system vehicle can contact the police department if a suspicious activity is noted. Other, less controversial programs include Weather Watch, which is part of the National Weather Service. It allows workers in the field to report bad weather to an official agency for broadcast. In areas that experience severe weather, the system may provide for assisting stranded motorists. These programs are advertised by the organization in many ways. Sometimes a special logo is painted on the truck or placed at the main office. In some cases, this information may be incorporated into a yearly bulletin or the utility bill. Larger systems may actually provide training to local contractors or smaller cities. Whether systems are large or small, notification hotlines for overflows and cave-ins are critical parts of public safety. Once again, these programs are advertised to the public as value-added services. Systems must be cautious not to infringe on the public sector or do so much that the public wonders who is actually operating the system. However, public service is an important perception that can be sustained by proactive public safety programming. In addition, some in-house training or safety programs can be presented in conjunction with programs from other agencies, schools, or the private sector. While the system benefits from the exposure, the employees also benefit. An additional advantage of these efforts is higher quality work: employees who feel they are part of an organization that is viewed in a positive light in the community tend to be better employees. While difficult to measure, the difference between organizations that provide special attention to safety and those that do not is easy for the public to see. It is perhaps during an organization's most troublesome periods that these efforts assist by establishing a positive perception in the minds of the public.

EMERGENCY PREPAREDNESS. Whether large or small, all wastewater collection systems need an emergency preparedness plan. The plan may be a simple understanding between the manager and employees, but it may also be a thick manual requiring training and shift assignments. For systems in Key West, a freeze-protection plan may not be a necessary part of emergency planning. For systems in the state of Washington, freeze protection is an essential component of operation. Table 1.1 provides a sample outline for an emergency preparedness plan based on a city in the southeast with a population just over 100 000. Each year, systems should review and update the departmental emergency preparedness plan and, if appropriate, coordinate the addition of new equipment and procedures.

EMPLOYEE HEALTH PROGRAMS. With regard to health programs, wastewater collection system operation requires special attention. Recommended immunization programs for service operators should be followed.

Table 1.1 Sample emergency preparedness plan outline.

Wastewater collection administration
Scope of plan
Emergency personnel assignments
Wastewater collection personnel
Organizational chart
Wastewater collection roster
Emergency support equipment
Communication equipment
Personal protective equipment
Rescue equipment
Equipment list and operator assignments
Emergency-related equipment list
Heavy equipment and primary operators
Wastewater collection emergency response plans
Hurricanes, severe storms, floods
Electrical outage
Bomb threat and terrorist activities
Explosion potential and hazardous material spills
Force main ruptures

Guidelines for employees exposed to wastewater in the eyes, mouth, ears, or through complete immersion should be accessible and well known to all employees. Health education must be provided to all employees. Shower facilities, special soaps, nail brushes, and other personal hygiene supplies should also be provided. Although the bloodborne pathogens standard was not initially intended to cover wastewater workers, they are now considered to be covered if they are exposed to raw blood or other sources of infectious materials emanating from hospitals or mortuaries. Employees exposed to sharps such as needles are also covered under the standard. In addition, workers who use self-contained breathing apparatuses or air-line systems have several health issues that relate to respiratory function and fitness testing. Simple approaches such as baseline health assessments before employment and uniform service with cleaning are becoming standard practices in the wastewater industry. Numerous U.S. Environmental Protection Agency studies list at least four families of disease in every sample of wastewater.

It is important that every system have standard polices in place to direct and control these issues. The organizational goal should be to reduce through engineering controls the employees' exposure to raw wastewater. Following that, specific procedures must be in place for items such as cutting force mains, removing check valves, and replacing air-release valves. Proper lockout/tagout procedures must be followed, including the procedures by which employees will bleed off pressure before opening or cutting any line, which may contain wastewater under pressure. It is inevitable that some workers will come into contact with wastewater, and managers must put in place systems and procedures that will provide appropriate responses to such exposures.

SAFETY REGULATIONS. The major piece of federal legislation addressing workplace health and safety, the Occupational Safety and Health Act of 1970, does not apply to publicly owned and operated wastewater facilities (WEF, 1990). According to the act, the definition of an employer does not include "the United States, any state or political subdivision of a state." Thus, the standards developed by the federal Occupational Safety and Health Administration (OSHA) do not always cover collection systems workers. However, many states have assumed responsibility for developing and enforcing health and safety standards that correspond to federal standards developed under the act by submitting regulatory plans to the secretary of labor. The approval of such plans requires that they provide for programs covering employees of the state and its political subdivisions. Many states have approved plans, and it is important that collection systems managers be aware of the requirements of their state occupational safety and health agencies' regulations. Further, regardless of the technical applicability of the Occupational Safety and Health Act, safety managers should understand and support its objectives and requirements. In worst-case situations where significant injuries or deaths have occurred, a third standard is often used during litigation: industry standards are often cited in wrongful death lawsuits. In such cases, comparisons are made between the facts of the case and what the industry recommends as a standard. It is important for managers to be aware of and, where possible, follow the established industry standards. Participation in professional organizations can provide insight into new issues. State member associations provide local information on state activities. By staying active in industry events, managers and supervisors can help their organizations by planning for changes rather than being overwhelmed by them.

SAFETY PROGRAM ORGANIZATION

The support of the top-level administrator of an organization is key to a successful safety program. In a smaller organization, this administrator may be the mayor. In others, the administrator may be the general manager of the utilities divisions. Many organizations use a resource person from their risk-management department to act as the interface between the top administrator and the operational area. In some cases, the administrator may work directly with the department or division managers.

Another approach is to have a designated safety officer coordinate the safety programming and training for each division. Again, this person works with the individual department managers but reports directly to the top administrator or general manager. An effective alternative is to have a department coordinator who works with a safety trainer. In this application, the department coordinator fulfills production-related and safety responsibilities.

Most wastewater collection departments are not large enough to warrant a full-time, in-house safety trainer. Although no absolute rule exists, a good service level for implementing a safety trainer is 200 employees. In many cases, this job responsibility would include all water and wastewater areas. This arrangement has proven to be effective because of the similar nature of many of the safety issues such as respiratory protection, lockout/tagout, and excavation. The types of responsibilities assigned to employees help determine the service level necessary to warrant a full-time safety officer. A typical approach is to identify an in-house employee who has demonstrated outstanding safety performance and reorganize his or her duties over a period of time. While additional training will be needed, this process can be broken down into individual and logical steps. This approach can be effective if basic communication skills are in place. Another important factor in the development of a working safety program is the establishment of levels and lines of responsibility. The ability to reward and discipline employees or stop work in progress is essential for any safety officer. In matters of safety, the authority of the safety officer equals that of the operational manager.

Ultimately, the front-line supervisor is in charge of safety on a daily basis. A key element in the success of safety programs is the working relationship between the safety officer and front-line supervisors. If properly supported, this partnership can result in an effective program.

Supervisors should require employees under their direction to fulfill their safety responsibilities. They should help train their employees in procedural safety, investigate accidents, and make recommendations for accident prevention. If a manager can convince field supervisors that safety is as important as production, the result is a better product and more efficient work. Supervisors will only put as much emphasis on safety as managers do. Managers should take the time during their next supervisor evaluation to discuss it at length.

SAFETY PROGRAM FINANCING

The budget process is an important part of the overall safety plan. Substantial resources are allocated to protective equipment for confined space entry, trenching, and other typical hazards associated with wastewater collection. Gas detectors, body harnesses, and tripods also require funding, as do consumable items such as disposable gloves, protective clothing, and antibacterial soaps. In addition, personnel costs and financing for incentive programs are included in the budget.

An effective safety plan includes options and alternatives. For example, the provision of a gas detector for every crew is a worthwhile goal, but not necessarily cost effective. Could a line locator or some other mobile person have one gas detector and perform testing for all crews on an interim basis? Often, organizations base cost effectiveness of equipment or materials on hazard exposure. At times, there are no safe alternatives to purchasing equipment; however, safety equipment can be purchased first, followed by the purchase of other safety-related equipment, as money becomes available. Whenever possible, costs can be shared. Items such as trench boxes have uses other than for wastewater collection, and other departments can share their costs.

Some safety items, such as barricades or gas-detection equipment, need to be maintained and eventually replaced. The budget must account for these items. Ideally, the safety budget would receive broad support from all parties involved in it. In addition, the cost of necessary service is an important consideration. For example, systems may calibrate gas detectors in house. Outside vendors provide this service, but there is a cost associated with it. A respiratory fitness test may be required for some employees. The costs of these and other health-related services required by regulation must be considered in the budgeting process.

MATERIAL RESOURCES

The makeup of the system determines the necessary safety tools, hardware, and equipment needed. This equipment is sometimes specified by regulation. Evaluation of the number of units and their replacement schedule requires an understanding of the working system.

One method for determining material requirements involves assessing immediate and long-term needs based on the overall plan. Both operation and training materials are considered in this process.

Materials typically needed for safety operations and training are as follow:

- Gas detectors,
- Tripods or nonentry rescue equipment,
- Full-body harness,
- Trench box,
- Manhole box,
- Disposable gloves,
- Rubber boots,
- Antibacterial soap,
- Scrub brushes,
- Safety glasses,
- Hard hats,
- Insulated probe rods,
- Traffic control equipment,
- Radios,
- Respirators,
- Self-contained breathing apparatus,
- Training facility,
- Audiovisual equipment,
- Cardiopulmonary resuscitation and first-aid equipment,
- Paper materials,
- Training videos and slides,
- Outside trainers,
- Simulators, and
- Computers.

RECORDKEEPING

Industries use two basic approaches to recordkeeping: they keep hard copies in departmental personnel folders or use a computerized database. Many computer-based programs are designed to help with this task. Department managers must have access to records and training histories at any time during a workday. Training coordinators need information on a broader scale, and a database allows them maximum access to information about appropriate groups of employees.

When a task or training module is completed by a training source outside the department, written documentation of that training is forwarded to the department for storage. When training is part of a larger program, such as a craft-training program, documentation is also recorded there. Many organizations provide a training manual for each employee so that they can track their own progress. The class instructor signs the manual when a student successfully completes a class. These student training records are only guides and are not recommended as official records.

While records typically are written after the training is done, they may also be the best planning resource. Everyone cannot be trained in everything; therefore, records provide a valuable tool for evaluating which employee needs which training next.

The centralized database approach provides cost-benefit advantages when large sections of employees have similar training needs such as for CPR and first-aid training. A trainer can use a database to monitor safety certification expiration dates and schedule a mixed training session for employees from several areas. This approach reduces productivity losses because entire departments are not shut down.

COMPUTER APPLICATION

In addition to offering database applications, computers greatly benefit safety training efforts in other ways. Forms, fliers, training announcements, diagrams, tests, student handouts, safety committee minutes, and other safety materials can be generated with a desktop computer.

The computer can also facilitate the provision of customized training packages in both a timely and cost-effective manner. Computer paper with a diploma-style border even allows for specialized certificates of achievement for in-house training modules.

Computers can also be used in conjunction with video equipment to develop video-based training modules. Using a digitizer or scanner, a computer can store line drawings and illustrations and reproduce them for use in program development. Newer, more advanced digital video equipment requires no tape or film. New cameras and editing systems are run through personal computers to allow the addition of graphics, sound, and titles.

NEW TECHNOLOGY

Atmosphere detectors have changed rapidly during the last few years. More models and features are now available. Sensors with 2-year warranties, internal boards with lifetime warranties, and calibrations that can be done in 50 seconds are just a few features of emerging technologies.

Nonentry rescue systems have also improved and now include fall protection and the adaptability of systems to work off multiple base units and on top of tanks. New lower cost harnesses featuring comfortable strap systems and additional positioning rings are now available. Several new styles of probe rods with insulated links are available to protect workers searching for cleanouts. Many new companies are offering antibacterial soaps to be used by wastewater collection system workers.

New regulations, technologies, and expectations are now faced by managers and workers. To ensure that safety programs are current and in compliance, collection facilities must stay in contact with OSHA and with state and local safety regulatory agencies. The new trenching regulation (Competent Person) and confined space regulation (Permit Required Confined Spaces) signify an increased emphasis on safety and liability.

REFERENCE

Water Environment Federation (1990) *Operation of Municipal Wastewater Treatment Plants*. Manual of Practice No. 11, Alexandria, Va.

SUGGESTED READINGS

National Institute of Occupational Safety and Health (1987) A Guide to Safety . . . Confined Spaces. Publication No. 87-113, Cincinnati, Ohio.

Water Environment Federation (1998) *Confined Space Entry.* Alexandria, Va.

Water Environment Federation (1994) *Safety and Health in Wastewater Systems.* Manual of Practice No. SM-1, Alexandria, Va.

Occupational Safety and Health Administration (1989) Standards-Excavation: Final Rule. Vol. 54, 29 CFR Part 1926, Washington, D.C., 45959.

U.S. Environmental Protection Agency (1987) *Operation and Maintenance of Wastewater Collections Systems.* Vol. 1, Washington, D.C.

U.S. Environmental Protection Agency (1982) *Other Wastewater Collections Systems.* Vol. 2, Washington, D.C.

U.S. Environmental Protection Agency (1980) *Wastewater Aerosols and Disease.* EPA-600/9-80-028, Washington, D.C.

Chapter 2
Design Considerations

This chapter considers design factors and how they affect operations and maintenance (O&M) for sanitary sewer systems. Many of the topics discussed in this chapter also pertain to storm and combined sewer systems. However, the inclusion of specific design issues for storm and combined sewers is beyond the intent of this manual; this information can be found in several of the suggested readings listed at the end of this chapter.

Criteria used in a particular sewer design typically are established in the preliminary stages of the project. It is advisable to seek input from maintenance managers, superintendents, geotechnical engineers, and construction resident observers during the development of design and routing criteria. People in these positions know where field problems are likely to be encountered; thus, they can offer valuable information in the early stages of the project.

Key design considerations that affect O&M include

- Line locations, grades, and depth;
- Route geography;
- Manhole spacing and size;
- Manhole channels;
- Influences of pumping stations;
- Odor control provisions;
- System instrumentation and controls;
- Induced turbulence (such as drop manholes);
- Sewer and manhole accessibility;
- Allowance for infiltration/inflow (I/I) capacity;
- Knowledge of system flow characteristics;
- Initial sewer flow rates compared to design flow rate;
- Degree of construction resident observation; and
- Proficiency of the sewer maintenance department.

Neither routine maintenance nor emergency service can be successful if maintenance personnel do not have appropriate access. One way to encourage the designer to include maintenance personnel in design criteria development and final design is to require the maintenance manager's "approval for maintenance" for the final design plans.

Ideally, collection systems should be designed to minimize total life-cycle costs: that is, the initial cost plus the present valve of O&M maintenance costs. This implies that a well-designed system can be easily and efficiently operated and maintained. However, some well-designed systems are never constructed as designed because of poor construction techniques, construction field changes, the absence of construction observation, and inadequate testing. Regardless of the reason, however, the majority of the collection systems will experience some deterioration and require general maintenance.

REGULATORY REQUIREMENTS

NATIONAL POLLUTANT DISCHARGE ELIMINATION SYSTEM PERMITS. National Pollutant Discharge Elimination System (NPDES) permit language contains standard references to collection systems in four permit conditions: noncompliance reporting, proper operation and maintenance, duty to mitigate discharges, and prohibition and defense provisions. Designers should review these requirements with the collection system owner and consider whether state or federal clarification of the permit requirements is necessary. For instance, are the noncompliance reporting requirements clear? Are there exceptional circumstances where prohibition and defense provisions need to be expanded or more specifically stated?

Currently, entities that operate and maintain a collection system, whether or not they have an NPDES permit, are responsible for any discharge to waters of the U.S. from their sewers. Municipalities or permit holders should be sure that local ordinances are in place that clearly establish the responsibilities and controls expected from flow-contributing private laterals and private satellite collection systems that are served and do not have an NPDES permit.

SYSTEM CONVEYANCE DESIGN RATES. Collection systems, or major hydrological sections of the collection system, should have an overall design rate that is accepted by state or federal regulatory agencies. Many communities develop this design rate (typically related to a design rain event) during the facility planning, sewer master planning, or watershed planning activities using historical and economical data. Design rates should include the planning period's wastewater flow projections and an I/I allowance that reflects the existing or targeted seasonal and rain event flow variances. Establishing an overall conveyance design rate will provide the basis for new construction and rehabilitation design criteria and help avoid hydraulic bottlenecks/constrictions in the system. Such bottlenecks can lead to sanitary or combined sewer overflows.

WET-WEATHER FACILITIES. Wet-weather facilities are facilities specially designed and constructed to manage or control flows associated with wet-weather conditions. Wet-weather facilities are intended to help protect the environment and public health of a community. These facilities may be designed to discharge under certain conditions upstream of a wastewater treatment plant's (WWTP's) headworks. Any facility designed with an anticipated discharge must be permitted. However, wet-weather facilities may be designed for off-line or in-line storage with no discharge.

Wet-weather facilities may include one or a combination of screening, primary sedimentation, in-line or off-line storage basins, chlorination, or dechlorination (site-specific stream conditions may not require storage). For example, the facilities will be designed to function according to site-specific, wet-weather conditions whereby the collection system wet-weather peak flows are diverted to the wet-weather facilities for storage. If the diverted flow rate or volume is greater than what the facility can handle, the excessive flow is discharged to a receiving stream after having gone through as much of the permitted process layout as possible. When wet-weather flows diminish, the stored wastewater is released back to the collection system.

Wet-weather facilities can help prevent the considerable O&M problems associated with basement backups and also help reduce, through their containment functions, the frequency of sanitary sewer overflows. Wet-weather facilities must undergo a comprehensive evaluation and justification process but can be the most reasonable and cost-effective alternative to larger conveyance systems that struggle to achieve adequate cleansing velocities during normal flow conditions.

DESIGN APPROVALS. Conveyance system design approval procedures vary between the states and localities. The designer should contact state, county, and city agencies to learn of any mandatory design standards or criteria that need to be addressed to receive regulatory approval.

ROUTING CONSIDERATIONS

ASSESSING ABOVEGROUND AND BELOWGROUND INTERFERENCE. Routing considerations will be different for gravity versus pressure sewers, but both should emphasize O&M accessibility in their design criteria. Many factors affect the final selection of a route. Important considerations for aboveground systems include

- Paved versus earthen surfaces;
- Surface water flooding potential;
- Proximity to residential and commercial buildings;
- Construction problems such as limited equipment movement or stability risks of adjacent structure;
- Traffic or recreational disruption;
- Biological, aquatic, or historical disruption; and
- Number of land parcels traversed and easement requirements.

Important considerations for belowground systems include

- Geology and soil conditions;
- Groundwater, river, or tidal influences;
- Other utility conflicts or influences (such as cathodic protection systems); and
- Depth of trench and safety requirements.

GEOTECHNICAL SURVEYS. Geotechnical surveys can reveal important information to the designer that will not only affect the design and construction costs, but also influence the long-term life of the conveyance system. Deterioration of the sewer and manhole materials can be accelerated if there are not adequate provisions for protection from corrosively aggressive soils. Surveys can reveal the extent of rock in an area and influence the feasibility of whether to install a gravity or pressure system. Surveys can also reveal soil expansion and stability conditions so that the designer can select materials and systems that will maintain water tightness while allowing for some lateral or longitudinal movement within the pipe and bedding zone.

PERMANENT ACCESS. Many existing sanitary sewer I/I and maintenance-related problems occur because a sewer is inaccessible and cannot be routinely evaluated. Permanent access roads or gate assemblies on unimproved land should be considered a part of new sewer construction projects where possible. The cost for constructing these access roads is not likely to decrease in the future, and most sewer contractors will have suitable materials and equipment already on hand to perform this task. However, construction of an access road across a wetland area should be carefully evaluated. The decision of whether to provide permanent access must be closely coordinated with permitting and other interested groups because wetlands by definition will vary in physical character. Some sensitive wetland areas may simply require too much time, effort, and cost to enable access. Alternatives that are environmentally sensitive, potentially cheaper, and less controversial should be evaluated.

The designer should work closely with the sewer system owner and legal staff to discuss the required temporary and permanent easement widths. If permanent widths are too narrow for the depth of the sewer, repair costs are likely to be significantly greater in the future because of extra trench shoring and accessibility restrictions. This is particularly true as the area becomes more developed and easements are encroached on.

The choice of initial routing location should allow for the transection of as few land parcels as possible. Designers and sewer system owners tend to underestimate the time needed to obtain easements. It is advisable to have all easements obtained (or at least the critical easements) before awarding the construction contract or beginning construction. A process and schedule for obtaining easements is as important to the overall project as developing construction documents.

PIPELINES AND JOINT MATERIALS

GRAVITY PIPELINES. Sizing Considerations. Gravity pipeline sizing should result from applying design criteria to system conveyance planning flow projections. In addition, there are many reference texts and manuals that provide concepts for sizing a gravity sewer. *Gravity Sanitary Sewer Design and Construction* (WPCF, 1982) is one example. However, selection of the pipe diameter is increasingly being influenced by O&M considerations, such as scouring velocities or provision of wet-weather, in-line storage.

Some communities have adopted sewer criteria that establish a minimum velocity of 0.76 to 0.91 m/s (2.5 to 3.0 ft/sec) rather than 0.61 m/s (2 ft/sec) at half or full pipe flow conditions. Other communities have oversized their interceptors to provide in-line storage during wet weather and incorporate O&M activities to induce cleansing velocities. Decisions regarding oversizing, and pipe and joint sizing in general, are being aided significantly by the increasing availability of affordable and technically reliable hydraulic modeling programs.

Materials. Proper sanitary sewer design includes careful evaluation of construction materials. Future maintenance and rehabilitation costs can be reduced by attending to factors that affect the life expectancy and use of the sewer.

Wastewater characteristics, site conditions, and sewer construction methods affect material selection. Factors to consider when evaluating a gravity piping system include

- Life expectancy and performance history;
- Availability of sizes required;
- Backfill and bedding material requirements;
- Availability and ease of installation of fittings or connections;
- Type of joint (fluid and pressure tightness and ease of assembly);
- Resistance to scour or abrasion;
- Resistance to acids, alkalis, gases, solvents, and other industrial wastes;
- Ease of handling and installation;
- Physical strength;
- Flow characteristics (friction coefficient);
- Cost of materials, handling, and installation; and
- Manufacturer testing and quality control program.

No single material is optimal for all conditions; selection typically is made based on the particular application under consideration. Different grades of materials may be selected for portions of a single project, but this practice typically is kept to a minimum unless a significant cost savings would result. The differential in the costs of various types of pipe tends to be a small part of the total project cost, and a more expensive pipe and manhole may reduce future O&M costs. *Gravity Sanitary Sewer Design and Construction* (WPCF, 1982) describes the more typically used materials and their advantages and disadvantages.

The basic pipe materials commercially available for gravity sewer construction and given in the references are

- Ductile iron,
- Concrete,

- Polyvinyl chloride,
- High-density polyethylene,
- Fiber glass, and
- Clay.

Ductile iron and concrete pipes typically have interior linings or exterior coatings to extend their life expectancy because of the aggressive and corrosive sewer environments to which they are subjected.

Backfill and Bedding. For the purposes of this discussion, backfill will include the zone of material around the pipe in a trench installation. The area the pipe rests on will be called the pipe *bedding*. Both the pipe backfill zone and the bedding area serve an important function in the pipe's ability to support the resultant load forces. The materials that can be used in the pipe zone and bedding area vary as widely as the soils in which they are buried. Therefore, it is important for the designer to coordinate the pipe's inherent strength characteristics with the type of backfill and bedding material that will be used during construction.

Attention to backfill and bedding requirements during design and construction can eliminate many typical O&M problems. Improper pipe backfill and bedding design or a lack of uniform bedding during construction can lead to pipe structural failures, differential joint settlement, pipeline sags, and admission of groundwater, soil, and roots. In addition to the pipe manufacture's design guidance literature, there are several good references the designer can review for a more detailed discussion of backfill and bedding design practices (for example, WPCF, 1982).

Designers should consider whether to include an allowance for future earthen loads when selecting a pipe and backfill combination. For instance, pipelines routed along undeveloped urban streams or ditches may experience in their design life several feet (metres) of fill material greater than the original earthen depth. This often happens when a developer wants to raise the foundation elevation of proposed new buildings above the flood elevation.

Deviation from designed trench or embankment widths during construction may result in an overload to the pipe. This is particularly true when a sewer pipe approaches and leaves a manhole. Designers should be aware that if a limited trench width is required for the backfill material specified, and if it is less than the width that is required to place a manhole in the trench, the potential exists for an overloaded pipe. Also, inadequate backfill compaction could cause differential settlement, overdeflection, or shearing, and ultimate pipe failure. An evaluation of the probability of differential pipe settlement versus structure settlement is part of the design phase.

Granular backfill fines can be scoured from a compacted backfill if the terrain grades are steep. The designer should consider installing a solid encasement around the pipe or a water barrier such as clay or other impervious material in the trench to impede subterranean flow along the pipe. Fines can migrate from the native soil in the trench walls into the granular backfill as well. This can cause a loss of lateral support and additional load bearing forces on the pipe. Flexible and semirigid pipe materials will be affected by the loss of lateral backfill support and possibly experience loads greater than what the designer had anticipated. One method to reduce the fines migration is to line or wrap the trench bedding and backfill zones with a geotextile material. This method is used also when the trench native soils are soft organic clays that might allow the granular material to move laterally when placed under load.

The O&M problems and pipe failures previously mentioned can also occur when the pipe diameter becomes sufficiently large that the installer has difficulty placing and compacting the backfill under the haunches of the pipe as required. This problem is more common for pipe diameters greater than 60.96 cm (24 in.). A flowable grout is an alternative; however, other issues such as the potential to float the pipe and chemical attack to metal pipes and appurtenances become considerations that must also be addressed.

Insufficient cover over a pipe can cause pipeline problems similar to those previously discussed. Operations and maintenance problems may develop faster with shallow cover because the failed or defective pipe is more susceptible to roots- or rain-induced I/I.

Grades. The composition of domestic wastewater has changed as society's needs and practices have changed. Many of society's convenience wastes, such as nonreturnable or disposable materials, find their way into sanitary sewers. Proper sewer grades help keep these wastes from settling.

Too often, sanitary sewer designs include long lengths of line laid at minimum grades, generally accepted as grades that produce 0.61-m/s (2-ft/sec) flows at half and full pipe diameters. A somewhat steeper grade that produces velocities of 0.76 to 0.91 m/s (2 .5 and 3 ft/sec) typically is preferable to reduce deposition and hydrogen sulfide generation.

On steep grades care must be given to the selection of pipe materials to prevent pipe wall erosion when velocities are greater than 1.8 m/s (6 ft/sec). At these velocities, the designer should consider using a drop manhole concept to reduce the flow velocities. Also, special attention to anchoring the pipe in the trench is important to prevent the pipe from creeping downhill. Anchors or retards should be coordinated with the appropriate backfill material. For instance, as previously mentioned, granular backfill in steep trenches may shift, causing the pipe to pull apart, particularly if percolating water velocities scour the backfill fines.

Sanitary sewers are designed for a future peak flow. For this reason, problems may be encountered in minimum-grade sewers during the early years of their design life. Low initial flows that do not approach design capacity can create maintenance problems and may require frequent line flushing and cleaning.

Infiltration and Inflow Allowance. Most sewer systems are never so watertight as on the day of their installation performance test. After a new sewer system is installed, it eventually develops defects or avenues for extraneous water to enter the system. Often, the older the system is, the greater the I/I; however, newer systems can sometimes exhibit as much or more I/I than older systems, depending on the quality of the construction and local conditions.

Design allowances for I/I can be developed for new pipe and collection systems from flow monitoring data collected on the existing sewer system or by benchmarking other similar sewer systems. Allowances for I/I should be coordinated with design standards that result from system conveyance planning design rates previously discussed.

PRESSURE PIPELINES. Types of Pressure Systems. Pressure systems can be either positive or negative. Most systems are positive-pressure, whereby a pump transfers energy to the wastewater that lifts the wastewater to the desired elevation at a specified rate. The most common positive-pressure system results when a pumping system receives wastewater from a gravity sewer or other pumping systems. Pipeline diameters for these systems typically are 10.2 cm (4 in.) or greater. System pressures typically range from a few metres to approximately 60 m (200 ft), with the norm being between 15 and 30 m (50 and 100 ft). These types of pumping stations are discussed in Chapter 4.

Another form of positive-pressure system is a low-pressure pumping system wherein a pump receives flow from a single facility or a small group of residential or commercial facilities. The type of low-pressure pump used is influenced by whether the pump receives the wastewater directly into the sump pump or from an existing settling compartment such as a septic tank. The designer has the basic option of using a centrifugal or positive-displacement type pump.

The pressures produced typically are less than 45.7 m (150 ft), and the discharge is to a common pumping station wet well or a gravity system. Pipeline diameters from low-pressure pumps are generally 3.8 cm (1.5 in.) and pump to a 5.08 to 10.16 cm (2 to 4 in.) diameter header pipe. These systems have application where geographical or topographical features would make a gravity sewer system cost prohibitive, such as in areas along lakes, bayous, or other waterfronts that either have high groundwater or variable terrain elevations. Low-pressure systems are also used in areas of rocky terrain or where housing is sparse.

Infiltration and inflow into both the traditional and low-pressure pumping systems is relatively minor compared to gravity sewer systems. Essentially, the only component of the pressure system that is subject to I/I entry is the sump or pump wet well.

A negative-pressure system (also referred to as vacuum sewer) produces a negative pressure within the pipeline between the residential or commercial facility and the pump. The wastewater is drawn to the pumping station where it discharges to a gravity sewer or another pumping system. A negative-pressure system typically has a pipeline diameter of 5.08 to 20.32 cm (2 to 8 in.). Such systems have the same applications as the low-pressure systems previously discussed. Negative-pressure systems also are subject to minor I/I because of the air tightness of the piping system. For additional detail on systems such as pressure, vacuum, and small-diameter gravity sewers, reference the U.S. Environmental Protection Agency's *Manual of Alternative Wastewater Collection Systems* (1991).

Sizing Considerations. Most piping systems function as a single-barrel pipeline; however, a dual or parallel pipeline is not uncommon where service redundancy is extremely important or where clearance conflicts might prevent a single-barrel pipeline from being installed. In either situation, sufficient internal minimum velocities of 0.61 m/s (2 ft/sec) and, preferably, 0.91 m/s (3 ft/sec) must be reached to resuspend organic and small, sand-sized solids that enter the pipeline. As with gravity sewers, the early years of a new installation may see flows that are at the low end of the design range. Therefore, the designer should evaluate whether minimum velocities will be reached daily. If not, it may be necessary to develop a weekly or more frequent O&M plan of controlling flows in the system to create enough flow to flush the pipeline. The extent of the flushing plan will depend on the length of the pipeline. As an alternative, it may be necessary to include a pipeline "pig" launcher and retrieval appurtenances in the system design. A "pig" is a plug-type insert that moves inside the pipeline and either uses the hydraulic force from increased fluid velocities forced between the pig and pipe interior wall or uses the hydraulic pressure to mechanically scrape or dislodge material from the pipe.

Maximum velocities for wastewater positive-pressure systems should not exceed 3.05 m/s (10 ft/sec) for normal, sustained flows. Wastewater conveys sand and grit that accelerates the inside deterioration or erosion of pipe coatings or wall thickness at bends or valves. Velocities from 0.91 to 1.82 m/s (3 to 6 ft/sec) are more desirable for sustained flows. The ability to stay within these ranges will depend on the flow range that the pipeline must convey.

Materials. Typical pipe materials commercially available for pressure sewer construction include

- Ductile iron,
- Concrete,
- Polyvinyl chloride,
- High-density polyethylene, and
- Steel.

The same evaluation considerations apply to pressure pipe materials as previously discussed in the selection for gravity pipe materials. However, there are additional O&M considerations when evaluating pressure pipes. Pressure pipes tend to be installed at shallower depths than gravity sewer pipes. Therefore, native soil character, surface contaminants that percolate through the soils, and stray currents from other protected structures can influence the exterior corrosion rate of the pressure pipe. Interior corrosion can be more aggressive in pressure pipes at high points of the pipe's profile and at the discharge to the atmosphere.

Pressure pipes, particularly nonmetallic pipe materials, should be installed with a system for recovering the location of the installed pipe. This can be done with accurate record drawings or by burying a wire, tape, or electric signal coupon above the pipe.

Backfill. Backfill has the same important structural support function as discussed for gravity sewer pipe. It is important to recognize that soil conditions can vary widely within a project, and it is typical to use local, excavated soil as backfill around the pipe. However, the soil characteristics influence the performance of thrust restraint for pressure systems and should therefore be closely evaluated with geotechnical engineers.

Earth loads typically are not as great an external load because of the shallow depths sometimes involved. However, live loads can become significant. Designers should thoroughly review the pipe manufacturer's backfill design suggestions when designing the pipe and bedding combination.

Restrained Joints. The design of thrust restraint for pressurized buried pipelines with push-on-type joints is important because of the damage that can result from joint separation. Pipe systems that have all joints restrained, either with restrained joints furnished by the pipe manufacturer or from a welded joint, may not require external thrust restraint.

Thrust occurs at changes in pipeline size or direction. Thrust forces can be hydrostatic, hydrodynamic, or thermal in origin; designers are responsible for evaluating the need to consider each type of force. The designer should review the pipe manufacturer's recommendations for restrained joints for the pipe considered and determine if the application of the restrained pipe length and method is appropriate for the proposed site conditions.

Resistance to thrust forces can be provided by concrete thrust blocking, concrete thrust collars, anchorage to a structure, or, indirectly, transmitting the forces to the adjacent straight pipe sections using pipe joints designed to resist shear and joint separation. The forces in the adjacent pipe are then transmitted to the soil by a combination of soil friction and lateral soil pressure.

Concrete thrust blocks depend on the allowable lateral soil pressure to develop the necessary resistive force. The allowable lateral soil pressure for the local conditions must be determined during the design and properly observed during construction. In adequate soils and for smaller diameter pipe, thrust blocks typically are the most economical method of resisting thrust. However, sometimes the soils are unable to resist the lateral force or dead weight force of the block. Also, there are times when space limitations or the probability of future excavations will dictate the need for an alternative restraint method.

Designers should also check for corrosion requirements to determine what protection is necessary for the thrust-restraint system.

Air Valves. Air valves should be an integral part of the pressure pipeline system. Air that enters or is entrapped in the pipe must be released. Because flowing water is subject to changing pressures and velocities, air is continually coming out of solution when the pipeline is in service. Unless air valves are properly located, this free air will accumulate at high points within the system. Serious flow restrictions or even blockage can result from the air pockets that form in the pipe. Also, these air pockets can contain corrosive vapors and should be vented to reduce interior deterioration.

Vacuum relief is equally important. Repeated vacuum conditions can damage the pipeline lining, cause failure of some materials because of stress cycling, or, if the water column separates, generate high surge pressures when the columns rejoin. Surge-pressure forces will be generated in the pipeline during situations such as pump startup, rapid shutdown, or sudden valve closure. The forces created can be substantial and need to be evaluated to determine their influence on the design. Whether surge suppression is accomplished through a surge-relief tank or air release or vacuum valves, provisions for their regular maintenance should be established.

Valve maintenance is especially important on wastewater pipelines. Corrosion-resistant materials for the entire assembly are often overlooked. Use of corrosion-resistive materials should be considered throughout the system—from the corporation stop through the shutoff valve and air valve internal parts. Otherwise, when, for example, a leak develops, maintenance personnel might not be able to close the shutoff valve or the corporation stop.

Sizing the air valve and selecting the appropriate valve type for the application are important factors. The designer should refer to the design guidance recommended in technical bulletins published by the manufacturer or included in design reference materials. These materials will also include recommendations for equipment and procedures for periodic valve cleaning. The designer should take into consideration these maintenance requirements and provide adequate access space for a maintenance worker to repair or remove a valve. This could include providing proper vault drainage, so as not to contaminate adjacent surface water or groundwater, and providing for odor control if the valves are close to residential or commercial establishments.

CORROSION CONTROL. Without proper consideration of corrosion control, the wastewater conveyance O&M staff will encounter continuous maintenance and repair problems from a deteriorating piping system.

This section will not discuss corrosion control methods in detail but will outline important considerations in the selection of steel, iron, and concrete pipe materials for wastewater conveyance applications because these are the most susceptible to corrosion. A corrosion engineer or manufacturer representative should be consulted for specific corrosion protection controls. In addition, the designer should not overlook the metallic components (such as fittings, bolts, bands, and valves) that are used with polyvinyl chloride, fiber glass, and high-density polyethylene conveyance systems.

Corrosion is a result of the reaction between the pipe material and its environment. Corrosion can occur on the interior of the pipe as a result of reactions with the wastewater. The exterior of a buried pipeline can corrode because of reaction with the soil. The factors that must be considered when evaluating materials are determined after field testing and evaluation of the environmental conditions.

Bonded pipe coating and linings serve to control corrosion. They isolate the pipeline from its environment. However, coatings or linings are not perfect, and additional techniques may be required to control corrosion. Portland cement concrete or mortar can also be used to protect steel in nondetrimental environments. This is the basis for protection of the steel components of concrete pressure pipe and for mortar-coated and lined steel and iron pipe.

Polyethylene encasement is an unbonded coating traditionally used to protect the exterior of iron pipe. This system uses polyethylene sheets that are loosely wrapped around the pipe and fittings. This barrier reduces the exterior corrosion by reducing exposure to oxygen, corrosive species, and stray currents. Polyethylene encasement may provide economical control from external corrosion in certain environments. However, it is important that the polyethylene encasement system be installed correctly to provide the necessary protection to the pipe.

Cathodic protection is useful for mitigating external corrosion of both metallic pipelines and concrete pressure pipe. Cathodic protection can be an external source of direct current voltage applied to the pipe so that the pipe becomes lower in electrochemical potential than the soil around it. Thus, the pipe becomes the cathode rather than an anode. "Sacrificial" anodes such as magnesium or zinc rods can also be attached to the metal pipeline and buried in the soil around the pipe. The magnesium or zinc rods corrode sooner than does the metal pipe because the electrochemical force developed between the metals causes the anode to lose the metal ions before the pipe does.

The application of cathodic protection to a coated metallic pipeline supplements the protection already afforded by coating systems. Cathodic protection systems can be quite complex and should be designed by corrosion specialists trained in assessing when and how to implement this method of control. Cathodically protected systems require regular O&M to read and monitor the status of the protection system.

REHABILITATION. Rehabilitation maintains the structural and operational integrity of the sewer system. The methods available to evaluate and rehabilitate the sewer system have become more sophisticated with the advancement of robotics and improved camera systems. *Existing Sewer Evaluation and Rehabilitation* (WEF, 1994) discusses this subject in more detail.

Both gravity and pressure pipelines will require repairs or replacement on segments of the pipe during the lifespan of the material. Many of the factors that create the need to repair or replace pipe have been previously discussed. Typically, the primary reason for repairing or replacing conveyance system pipe is related to structural or I/I problems.

Structural. The need to make structural repairs typically becomes evident either through a sudden pipe failure or through field inspections such as internal pipe closed-circuit television inspections or pipe wall coupon sampling. In each case, an exact cause of the problem needs to be determined so that the corrective action is effective. Otherwise, the corrective action may have to be significantly overdesigned to account for this imprecise assessment and all of the possible or speculative causes of the problem.

Structural repairs typically consist of replacing the section of piping system with the same material. It would be advisable to use a granular pipe bedding material in conjunction with the pipe to provide a stable, uniform pipe support.

The designer should also consider using the pipe manufacturer's recommended replacement couplings and suggestions. If a section of pipe is replaced, the internal pipe diameter of the spliced-in pipe should match the internal diameter of the adjoining pipe sections. Otherwise, the joint between the old and new pipe section could become a point where solids and debris become lodged.

Besides using new pipe to make the structural repair, repairs for both pressure and gravity pipelines can be done with clamps or sleeves around the pipe exterior. Repairs also can be made from the inside of the pipe using a pipe liner material either for long stretches or for lengths of only several metres. For the longer pipe segments rehabilitation systems include slip-lining, swaged and rolled-down pipes, segmented linings, cured-in-place (CIPP) systems, and fold-and-form pipe (FFP) systems. The systems have become increasing competitive and beneficial because of their less disruptive trenchless technology features. Both CIPP and FFP systems have application for point structural repairs.

A critical principle to consider in the structural design of liners is that the load the liner must support is determined by the degree of degradation of the original pipe. Design conditions for liner structural design typically fall into two primary categories: partially deteriorated and fully deteriorated. There are many materials and systems for both the external and the internal materials that are commercially available. These materials and systems may be advertised in national trade organization publications.

Pipe bursting is a form of pipe replacement that is gaining interest. This system breaks the original host pipe, and a new HDPE pipe is pulled into its place. Pipe upsizing of one to two diameters is possible under certain conditions. Both pipe bursting and liner systems require that individual service connections be reestablished either by using robotics from the inside of the pipe or by excavating to expose and reconnect the service connection.

INFILTRATION AND INFLOW. Infiltration and inflow is a problem associated with gravity piping systems. The term refers to extraneous flow that enters underground sewer pipes and imposes an undesirable hydraulic load on the sewer system. Infiltration sources include groundwater infiltration, which may be prevalent year-round, and rainfall dependent infiltration and inflow, which is present during and shortly after rain events. Inflow is the extraneous flow that enters the sewers quickly through direct connections to the sewers. These direct connections are primarily through manhole covers, surface or yard drains, roof downspouts, sump pumps, or service connection cleanouts with missing caps. However, indirect inflow can enter underground sewer pipes. Stormwater exiting the storm sewer system through damaged pipes, nonwatertight joints, and unsealed lift holes may follow nearby sanitary sewer trenches and enter through damaged pipe and nonwatertight joints.

Rehabilitation factors associated with infiltration are different than rehabilitation factors associated with structurally damaged pipe. The extent of structural rehabilitation problems typically can be defined by television inspection of the sewer. However, the required rehabilitation to control groundwater infiltration is influenced by the migrating capabilities of groundwater. Therefore, infiltration rehabilitation work may include sealing or making each sewer and service connection joint watertight the entire length of the sewer within a groundwater zone of influence. Rehabilitation might include chemically grouting each joint (or at least pressure testing each joint) or relining the original sewer with one of the many commercially available lining materials.

MANHOLES

Manholes are installed in sanitary sewers to provide access for routine or emergency maintenance. They are sometimes rendered inaccessible because of location, negligence (for example, covered by pavement), or design deficiencies. Locating sewer lines in street or alley rights-of-way is preferable to the use of easements, where obstructions such as utility poles, meters, junction boxes, fences, retaining walls, trees, shrubs, or utility buildings can restrict access.

Manholes typically are placed at each change in direction so that sewers are straight between manholes. This practice has resulted from the following considerations: straight sewer layout and construction is easier; accurate record drawings have not often been kept to allow relocation of curved sewers; and pulling closed-circuit television and cleaning equipment in curved sewers is more prone to operational problems and pipe damage. Curved sanitary sewers are more prevalent in older systems compared to current practice even though a properly designed curved sewer will convey wastewater just as effectively as straight sewers and can cost less than a straight-laid sewer. However, some communities have adopted the practice of installing a manhole immediately upstream, downstream, or at both ends of the curved portion of the sewer if additional construction work is done in the vicinity of a curved sanitary sewer.

Ideally, manholes are constructed large enough to provide space for a person to work with cleaning equipment, with a shelf wide enough to provide safe footing and able to support a ladder. Manhole walls can be sealed from groundwater infiltration by using mortared joints, adequate flexible sealer, or rubber O-ring gaskets for precast-section manholes. It is reasonable to consider a combination of joint sealing methods because concrete casting tolerances may not produce tight-fitting tongue-and-groove precast joints.

Manholes typically have an inside diameter of at least 1.2 m (4 ft). Larger pipelines need correspondingly larger manholes to provide an adequate shelf on which to work and a wall surface sufficient for a solid, watertight connection. Eccentric manhole shafts, which have the steps along the straight, vertical side, are easier and less dangerous to enter. To reduce problems with some maintenance equipment, the straight side is sometimes over the outlet pipe. Many agencies require construction of concentric manholes, which permit cleaning regardless of orientation. Because of safety considerations, some agencies have eliminated manhole steps and now use ladders or hoists.

SPACING. Manholes typically are spaced at intervals of not more than 120 m (400 ft) for small sanitary sewers. Spacing is often specified by local regulation. With the advent of modern cleaning equipment, some agencies have considered lengthening this maximum interval. Although some new cleaning equipment is able to clean up to approximately 300 m (1 000 ft), other maintenance activities, such as bypassing a section or certain rehabilitation activities, can be difficult or impossible at that distance. Where there are truck access limitations (such as in a backyard or parkway), the distance from the access point to the manhole needs to be considered in determining the manhole spacing. Intervals of 150 to 300 m (500 to 1 000 ft) are typical for large sanitary sewers of 0.75 m (30 in.) diameter or greater.

A manhole is also necessary at each change of grade, invert elevation, or pipe size; at each intersection of two or more sanitary sewers; at the terminal end of each sewer line; and at major service connections (such as industrial facilities and schools). Avoid placing manholes at busy street intersections.

The practice of replacing the last manhole with a lamphole or vertical riser is not recommended. When sanitary sewers must be located along a rear property line, the sewer should be extended to a street right-of-way or an access easement, even if this requires the construction of an additional dry sewer. This ensures access for the maintenance crews.

MATERIALS. New Construction. Most new manholes are made of precast concrete. Other materials such as fiber glass and high-density polyethylene are also used in environments that are detrimental to concrete. Some communities require coal tar epoxy, elastomeric epoxy, or polyvinyl chloride liners in the manholes that are to be located on interceptors greater than a specified diameter. The walls of new manholes typically are heavily pitted and need to be filled with a cement layer before application of an epoxy liner. Otherwise, corrosive gases will penetrate the pinholes that exist, corrode the area behind the liner, and eventually delaminate the liner. Precast concrete manholes that are located in the upper reaches of collection systems or where hydrogen sulfide generation is minimal will provide years of good service.

Rehabilitation. Manhole rehabilitation typically results from investigations initiated by I/I problems in the collection system. Many communities like to rehabilitate not only their I/I defects but also their structural defects under the same effort. Therefore, a combination of structural, waterproofing, or corrosion control methods may be necessary at the same manhole. There are product manufacturers that advertise a product to meet all three requirements. The broad categories of manhole wall rehabilitation methods are chemical grouting, epoxy linings, waterproofing, sealing, coatings, structural linings, and cement grouting.

CHANNELS. Smooth, continuous manhole channels efficiently move wastewater from one or more inlet lines to the outlet lines. Properly designed and constructed side or angle inlets prevent deposition of solids at the transition point.

Wherever a change in pipe diameter occurs, the pipe crowns or water surfaces are located at the same elevation to reduce flow resistance. Where there is a change in direction without a change in diameter at a manhole, a drop in elevation is allowed through the manhole. The following drops are recommended: a 0- to 45-deg angle change over a distance of 3 cm (0.1 ft) and a 46- to 90-deg angle change over a distance of 6 cm (0.2 ft).

Inlet pipes that are too high to enter the manhole at the bottom (more than 31 cm [1 ft]) can enter through a drop connection. However, every effort should be made to avoid sudden drops in flow. Sudden changes in flow direction will cause gases such as hydrogen sulfide to be released out of solution.

Constructing shelf areas so that they drain to the channel at a slope of at least 1:12 reduces deposition. Because of careless construction, a channel of good hydraulic properties, though important, is not frequently created. The channel should be, as far as is possible, a smooth continuation of the pipe. To this end, the pipe is sometimes laid through the manhole with the top half removed to provide the channel (WPCF, 1982). The channel walls should be formed or shaped to only three-fourths the full height of the crown along interceptor sewers and other manholes with high velocities. This practice will help reduce deposition of solids on the shelf and facilitate smooth flow hydraulics through the manhole. Making the channel wall the full height of the sewer for most 0.2 m (8 in.) diameter sewers increases the difficulty of sanitary sewer investigations and maintenance tasks such as

- Physical inspection of pipes by lamping,
- Insertion of television inspection cameras,
- Deployment of cleaning equipment,
- Insertion of sewer plugs, and
- Installment of some flow measurement equipment.

COVERS. Manhole frames and covers are constructed of heavy cast iron or ductile iron. They need to be approximately 61 cm (24 in.) in diameter to allow easy access. Machining the horizontal bearing surfaces between the cover and frame reduces rocking and provides a tighter fit, especially if a rubber gasket is grooved into the frame or lid.

In flood plains, flood-prone areas, or areas subjected to ponding or sheet flow runoff, manhole covers should be watertight (with or without bolts). These covers and rubber gaskets virtually eliminate inflow. Lifting holes or bars in the lid are designed to be nonpenetrating to reduce surface water inflow. Removal methods that do not require lifting holes are preferred; however, if lifting holes are necessary, there should be no more than two. Alternatives to replacing an existing hole-penetrating cover are either hole plugs or frame and cover inflow inserts (dish). The inserts stop or impede the inflow from entering the system through the cover penetrations. The inserts are made of plastic, polyethylene, or metal materials.

Where manholes are located in poor soils, shifting sands, or soft asphalt pavement on a hot day, it is common to install a concrete ring that extends approximately 0.3 m (1 ft) around the top of the manhole frame and cover. This allows maintenance crews to have a firm footing for a confined space tripod and for lifting bars. A secondary benefit of the concrete ring is that it makes the manhole easier to locate both in drive-by searches and in aerial photographs.

Adjustment rings between the cone or slab and the rim casting may be appropriate in manholes located in streets and alleys. These rings facilitate adjustment for regrading projects. Take care to waterproof and seal the joints between the manhole frame, adjustment rings, and cone because this is often the area where inflow enters the manhole.

In cold climates or where traffic loads are heavy, adjustment rings may fail and allow I/I or loss of backfill. Rings of high-quality concrete may be used. Flexible frame-to-cone seals that permit movement from frost action or traffic impact are a viable alternative.

STEPS. Corroded or poorly grouted manhole steps (rungs) present a safety hazard. These steps can break off underfoot, possibly causing a serious or fatal fall. Safe manhole steps are made of heavy cast iron, stainless steel, fiber glass, wrought iron, or some other corrosion-resistant material. Bent steel or reinforcing bars are never recommended. Safety concerns require that steps be tested frequently for proper anchorage and removed or replaced if loose. Some agencies are dispensing with steps in new manholes and equipping maintenance vehicles with lightweight aluminum or fiber-glass extension ladders or portable hoisting frames and harnesses.

DROP MANHOLES. Drop manholes typically are considered drop manholes when the vertical distance between the influent pipe and the effluent pipe is greater than 0.3 m (1 ft). Drop manholes that have a drop greater than approximately 0.6 m (2 ft) should have two influent pipes. The top pipe enters the manhole on the same grade as the upstream sewer pipe. The bottom pipe has the same vertical alignment as the upper pipe but comes into the manhole at the invert of the manhole.

Drop manholes typically are considered when it is not cost effective to make an influent line match the effluent line, for instance, when the construction cost savings of installing the sewer pipe at the extra depth exceed the cost of the extra piping and workmanship of the drop connection at the manhole. However, because drop manholes can create other costs associated with materials corrosion and maintenance problems, other approaches are often used. For more information, see *Gravity Sanitary Sewer Design and Construction* (WPCF, 1982).

CONNECTION BETWEEN MANHOLE AND SANITARY SEWER. The design should anticipate the possibility of differential settling of the manhole and the sanitary sewer pipe. Design and construction precautions are warranted where the manhole and sewer connect. Shear cracks or stresses in a sewer pipe can occur at or immediately outside the manhole. The use of flexible pipe-seal connections integral to the manhole wall is a method typically specified in design to avoid buildup of stress forces in the pipe. If flexible pipe-seal connections are not applicable or used, a flexible pipe joint at or within approximately 1 m (3.3 ft) of the outside manhole wall decreases the likelihood of pipe damage.

Designers should allow for the change in trench width as the pipe enters and leaves the manhole. Also, the contractor should properly place, compact, and support the pipe with an appropriately graded granular bedding and backfill material. Otherwise, the sewer pipe may develop shear stresses and possibly fail.

Corrosion problems may be anticipated where a force main discharges into a manhole. Hydrogen sulfide (generated principally by the bacterial reduction of sulfate in wastewater) is released, oxidized, and condensed on exposed surfaces as sulfuric acid. This process can cause extreme damage to cement, carbonaceous aggregate, asbestos-cement, unlined ductile iron, and steel. Limiting turbulence at these juncture points will reduce the release of hydrogen sulfide gas.

MANHOLE AND PIPELINE TESTING

NEW CONSTRUCTION. Emphasis on eliminating I/I from the collection system has spurred the enforcement of performance tests that allow only relatively minor amounts of infiltration and exfiltration. Hydrostatic tests typically are used for pressure pipelines, manholes, and to a lesser extent small-diameter gravity sewers. Pressures for pipelines are established based on design requirements of the pipeline system and should represent the greatest pressure to be generated in the pipe. Pressures for manholes are static pressure only; the manhole is filled with water and the rate of exfiltration is measured.

Air testing typically is used for gravity sewers and manholes. However, manhole pressure tests are negative-pressure or vacuum at approximately −14 to 21 kPa (−2 to 3 psi). Pressure increases over time are measured against acceptable rates.

In addition to air or hydraulic tests, it is a good practice to pull a mandrel through the installed pipe or to require a video inspection of the installed pipe to note any observed leakage or defects. Mandrels are used to determine whether an installed pipe has deflected more than accepted by the designer. Mandrel tests typically are used with plastic pipe installations because they will deflect easily if not properly backfilled. Two basic types of mandrels are available. One is a rigid mandrel constructed with a fixed deflection that if exceeded will hedge in the pipe and not advance in the pulled direction. The other is a mandrel with telescoping prongs that are depressed by a deflected pipe and when removed from the pipe are measured against the designer's allowable deflection criteria.

Excessive deflection reduces flow capacity and can worsen over time. Television inspection for new construction can reveal manhole and pipe leaks or defects that may exist but that do not cause an air or hydraulic test failure. Also, television inspections can often reveal line sags or sediment or debris left in the pipe after construction.

Some communities require each manhole and pipe to be tested or television inspected. Others establish a minimum percent to be tested or televised with provisions for testing or inspecting greater numbers if specified criteria are not met. An advantage of television inspecting the entire system before acceptance (and before paving above the pipe) is that it not only confirms that the system is installed correctly, but also provides a documentation for future comparisons.

REHABILITATION. The same tests for new pipelines and manholes apply to rehabilitated pipelines and manholes. However, rehabilitation of manholes should include additional tests typically not required for new construction (except for the previous cases). These tests are a surface density test (hammer test), an adhesion test, and a pH test.

Surface preparation for manhole rehabilitation is a key factor in achieving adhesion of the structural or waterproofing liner to the manhole substrate surface. Surface density tests, pH tests, and petrographic tests (ASTM C856-83) can provide important information about the condition of the substrate material. Petrographic tests performed by a qualified person can reveal the concrete condition and the thoroughness of surface cleaning or removal activities. The adhesion test is the indicator test that ensures that the complete liner is properly applied.

INVERTED SIPHONS

An inverted siphon is a section of sewer that, because of a large obstruction such as a river, depressed freeway, subway, or utility line, must be installed below the hydraulic gradient and thus operates under positive pressure. Most inverted siphons (or depressed sewers) contain at least two pipes (barrels): a smaller pipe for low flows to provide greater velocity for scouring and a larger barrel to carry peak or future flows. To eliminate overflow or backup conditions, some inverted siphons include another pipe capable of carrying the entire flow in an emergency.

Inverted siphons typically have maintenance access structures at each end. Gates or stop plates at the inlet structure allow the diversion of all flow through any one barrel for scouring. Flap gates held open by a chain and accessible from the top of the manhole have been used successfully in small siphons. Working space is allotted in the inlet structure to launch a cleaning "pig" and in the outlet structure for pig recovery. To allow passage of cleaning equipment, smooth curves of adequate radius (less than 22 deg) are preferred. Inverted siphons have been successfully constructed with various types of pipe, including ductile iron, concrete, and polyethylene.

Odor problems tend to occur at the inlet chamber because air is released by air movement induced by the wastewater flow and the lack of a free-water surface through the siphon. Positive pressure develops in the sewer upstream of a siphon because of the downstream movement of air induced by the wastewater flow. Air tends to exhaust from the manhole at the siphon inlet and can cause odor problems. The problems typically are cured with air lines, or "air jumpers."

The inlet chamber has a lateral overflow weir and sometimes a sump to trap grit and heavy objects and to prevent clogging of the barrels. Screens may be installed if the added expense of periodic cleaning is warranted. Inlet and outlet structures both need tamper-proof manhole covers.

Both the inlet and outlet structures have prime environments for hydrogen sulfide release, which results in an aggressive corrosion attack on concrete and iron materials. Corrosion control is vital for these facilities if a reasonable life expectancy is desired.

SANITARY SEWERS ABOVE GRADE

Sanitary sewers are sometimes placed over an obstacle where topography, width of crossing, and flood elevations and velocities allow. They are installed or hung on bridges, supported on piers, suspended by cable spans, or fastened to a structural beam. Pressure-type joints, insulation, and jacketing prevent leakage and protect the pipe against freezing, although abovegrade sewers are rarely used in cold areas.

STREAM CROSSINGS

Sanitary sewers that transect a stream's flow and are installed using traditional excavation techniques are often installed with the top of the pipe located several metres beneath waterway inverts. This typically is done by necessity because there is minimum slope available, forcing the pipe to be as shallow as possible. These crossings are subject to erosion and washout because of their proximity to the stream's invert. Crossing pipe can be encased in concrete or placed in another pipe to protect against scour, impact, and potential inflow. Pipe joints designed specifically for crossing bodies of water are also available. If the crossing is complex, expected to experience considerable construction and installation risks, or simply a costly project, alternative crossing concepts such as microtunneling or directional drilling should then be evaluated.

On either side of a navigable stream (or lake), crossing signs should be posted that are visible from the land and water. The signs should alert the public to the presence of the submerged pipe so that activities such as dredging, geotechnical explorations, and boat anchoring do not penetrate or shift the pipe's position and make it more vulnerable to additional abuse. A contact phone number and any other special notes should be legible from a reasonable distance as determined by the specific site conditions.

Substantial problems may be encountered where sanitary sewers span a watercourse, particularly when support piers are required. Brush and debris invariably jam against the pipe, sometimes causing a rupture and discharge of untreated wastewater to the stream. Bridged crossings ideally are placed above the maximum flood level. Stabilizing stream banks helps to reduce erosion maintenance. Manhole structures are protected or physically located to reduce washout by bank erosion.

SERVICE CONNECTIONS AND DISCONNECTIONS

Service connections are also referred to as house connections, building sewers, service lines, or service laterals. They connect the customer's facility (building) to the public sewer system. Service connection requirements for each community vary as to the criteria that have to be designed and installed.

Attention to service connections has greatly increased over the past 10 to 15 years because of the recognized difficulties with I/I. Because the length of service connection piping can equal or exceed the length of public property sewers, their contributing impact on wet-weather flows can be significant. This fact has recently prompted a change to the long-held cultural restraint of investing public money on private service connections. As for service connections, public money may be more effectively spent to remove their defects. Regardless of the approach to correcting service connection defects, designers of sewers and WWTPs should include realistic flow allowances for I/I from service connections.

Optimum sanitary sewer design includes properly located and securely plugged "Y" and "T" connections for future service connections. It is typical for the public sewer to be installed and then the service Y or T connections installed with a section of service pipe plugged at the end and extended to the private property boundary. Close attention needs to be given to designing a tightly plugged pipe; otherwise, the service connection can become a significant source of I/I and sediment. Inexpensive mechanical plugs are a good concept to consider. Also, accurate location records help make Y and T connections easily recoverable when the plumber wants to complete the building connection or perform repairs on the Y or T connection.

Sometimes, service connections are not available on the sewer at the desired location. Service connection installation via the "break-in" or "hammer tap" methods can cause pipe failure and line blockage and makes pipes vulnerable to infiltration and root intrusion. Local regulations typically prohibit break-in service connections and require machine tapping and saddle-style or other suitable factory-fabricated connections to be made or approved by sewer authority personnel.

Old structures may be razed and the excavated land backfilled during property development. Typically, the sanitary sewer service is not properly disconnected, and the new construction design on the site requires a new service connection. Several generations of site reuse may result in multiple service connections. Inactive connections are a source of infiltration, insect and rat infestation, and other possible maintenance problems. Therefore, building permit and design regulations may often require that abandoned services be disconnected and permanently plugged at the main, if possible. Again, attention to having a watertight plug and an accurate location typically will restrict the potential problems that could otherwise result.

References

U.S. Environmental Protection Agency (1991) *Manual of Alternative Wastewater Collection Systems*. EPA-625/1-91-024, Washington, D.C.

Water Environment Federation (1994) *Existing Sewer Evaluation and Rehabilitation.* Manual of Practice No. FD-6, Alexandria, Va.

Water Pollution Control Federation (1982) *Gravity Sanitary Sewer Design and Construction*. Manual of Practice No. FD-5, Alexandria, Va.; Manuals and Reports on Engineering Practice No. 60, Am. Soc. Civ. Eng., Washington, D.C.

Chapter 3
System Maintenance and Management

Efficient collection systems operation requires good recordkeeping and planning, a competent workforce, and the purchasing of equipment appropriate to the system. These elements depend on system size, topography, complexity, age, and anticipated growth. With advances in monitoring equipment, computerization, maintenance equipment, and safety awareness, it is no longer sufficient to apply reactive management approaches to collection systems procedures. Throughout this manual, emphasis has been given to meaningful inspections (both for new installations and overall system conditions), the use of advances in computer-assisted generation of information and work orders, and comprehensive planning of system maintenance. This chapter focuses on establishing and operating a maintenance system and emphasizes preventive maintenance. The use of contractors in collection systems maintenance, beyond the experience or capabilities of an agency, should also be a consideration when assessing needed work. See Chapter 6 for a more detailed examination of outside contracting.

Goals

Goals determine what courses of action are needed and help set an operating plan in motion. Goals define the purpose and anticipated results of system operation policy:

- Prevent public health hazards;
- Reduce inconvenience and damage by responsibly handling service interruptions;
- Protect municipal investment by increasing the useful life and capacities of the system and parts;
- Identify and remedy design and construction deficiencies;
- Use operating funds efficiently;
- Convey wastewater with minimum inflow, infiltration, and exfiltration;
- Identify and remedy design and construction deficiencies;
- Prevent excessive claims and legal fees related to backup by providing immediate, concerned, and efficient service to all emergency calls; and
- Perform all activities safely and avoid injury.

Principles to follow for achieving these goals and rules governing the use of collection systems include

- Executing a basic routine preventive maintenance plan designed to prevent service interruption and protect capital investment;
- Immediately investigating all complaints and promptly correcting faulty conditions;

- Routinely inspecting for physical damage, repairing any damage, and eliminating the cause;
- Considering personnel safety in all operations;
- Recognizing public ownership and performing all functions courteously, efficiently, and professionally; and
- Implementing a community relations program to prevent abnormal and improper use.

Rules governing the use of wastewater collection systems include the following:

- All waterborne wastes must be discharged to a public collection system wherever such service is available, consistent with public health requirements and applicable regulatory provisions.
- No discharge contrary to the intended collection systems function so as to cause an inhibitory or toxic effect on the treatment process will be permitted.
- No use of the collection systems that may physically damage the system, cause an explosive condition, be harmful to human life, or interfere with normal maintenance and operation will be permitted.
- No connection to the system will be permitted without agency approval.
- No discharge of stormwater from roof, street yard drainage, basement pumps, or any uncontaminated water will be permitted, except where the system is designed for combined flow.

ORGANIZATION

Agencies are designed to perform and monitor operation and maintenance (O&M), administer budgeted monies, record and retain system information, and plan for future repairs and growth. They are also responsible for providing emergency service and public information.

An organization's size depends on the size of the collection system. Determination of staff requirements for a collection system requires a working knowledge of the system, including variables such as

- Level of service expected and funded;
- Miles of pipeline, subdivided into public and private sector systems;
- Age of the system;
- Construction materials used;
- Construction standards used;
- Agency responsibilities;
- Extent of contractor use;

- Equipment types;
- Work site accessibility;
- Travel distance;
- System area topography;
- Collection systems service contract obligations; and
- Customer classes and nature of industries.

No single formula can be used to determine the staffing needs of a particular system. The number and type of operating personnel vary by agency and change with job assignments, responsibilities, and number of satellite yards or maintenance districts. Typical staffing needs for a collection system serving a population of 5 000 are going to differ from those of a system serving 500 000 in that personnel in smaller agencies typically are cross trained for several operational duties in both the treatment and collection systems. As an example, in a small system serving 50 000 people and representing approximately 200 miles of sewer lines, three full-time and two part-time field maintenance people may be all that are required. In general terms, fieldwork typically requires two people to operate cleaning or video inspection equipment or perform routine inspections of the system.

Communities of fewer than 150 000 people typically only have one yard or maintenance district. Agencies serving more than 150 000 people, but fewer than 300 000 people, typically are divided into two maintenance districts in some cases served from separate maintenance yards. Agencies serving more than 400 000 people can have as many as three maintenance yards or districts. Staff size also depends on unit operations such as maintenance of the collection system and appurtenances (not including service lines), maintenance of stormwater lines (some agencies have separate crews for this), maintenance of pumping stations, and inspection of new lines and connections.

The function of the collection system is to collect, contain, and transport wastewater to a point of treatment. Ideally, a collection system keeps wastewater in and surface water and groundwater out, providing an uninterrupted flow. Staff requirements can be substantially affected by a routine preventive maintenance program; conversely, time spent on preventive maintenance reduces time spent on emergency situations that can drain financial resources.

All personnel, from the manager to the entry-level employee, are most effective when they clearly understand their responsibilities. Carefully written standard position descriptions and task instructions contribute to this goal. Figure 3.1 shows a suggested chain of command and career development ladder for a large organization (Chapter 4 addresses specific staffing requirements for pumping stations).

ADMINISTRATIVE STAFF. Administration management positions can include a superintendent, an assistant superintendent, and, in large agencies, a director or executive director. Individuals in these positions typically are

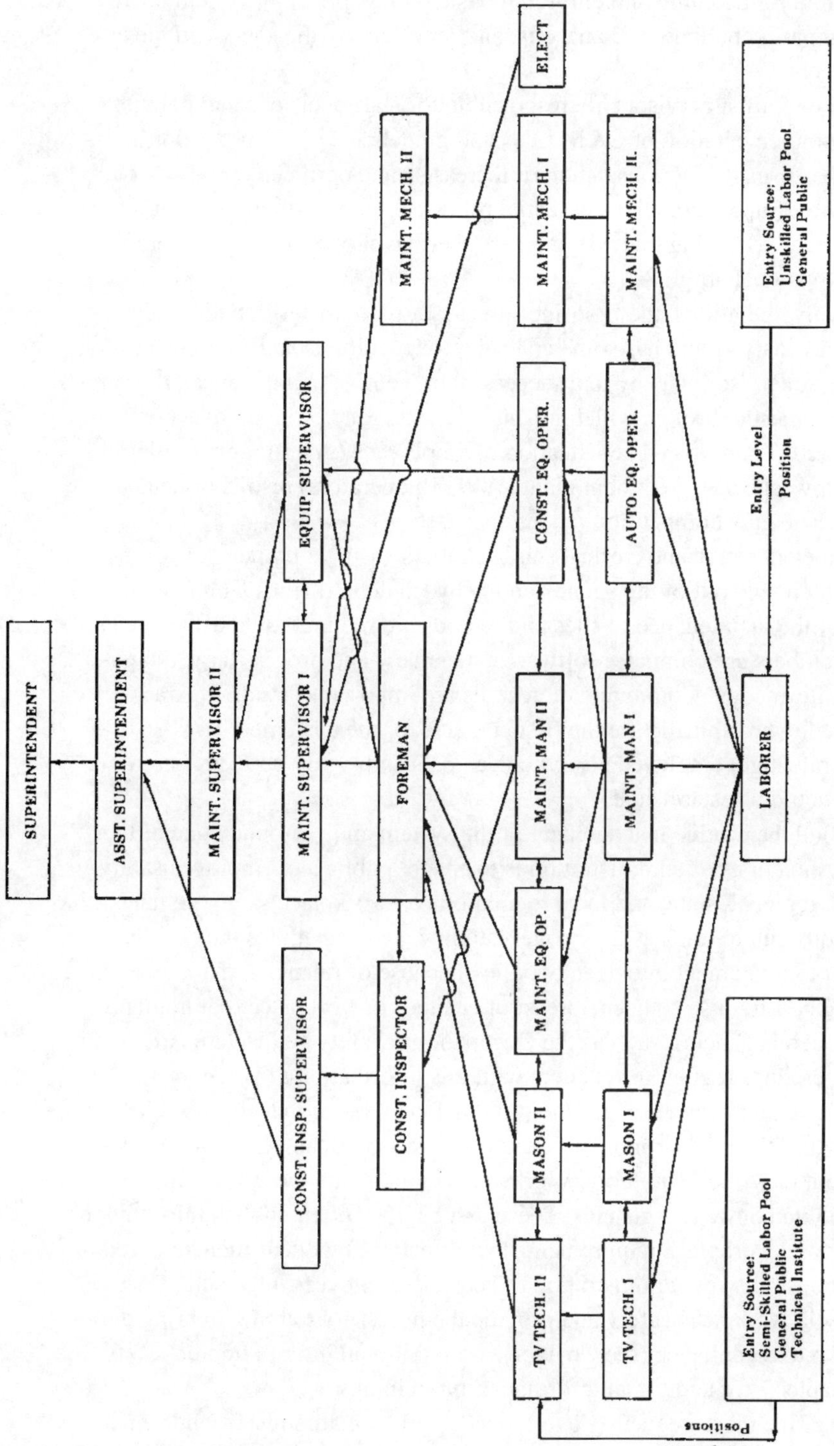

Figure 3.1 Typical wastewater collection system career lattice.

responsible for deciding and enforcing basic policy, planning system operations, preparing budgets and analysis, and representing the agency in the city government.

Field or staff supervisors are responsible for staff decisions and policies such as implementation of O&M rules and guidelines, personnel actions, employee performance appraisals, merit increases, and enforcement of safety procedures. Supervisors have essential roles in employee motivation and thus should receive training that will increase their level of expertise in operation and motivational abilities.

In many operations, the first-line supervisor participates in budget preparation and analysis and has some control of expenditures and personnel training. The supervisor is a key to a successful preventive maintenance program and is the liaison through which inventory control and program data are transmitted. Training and certification of employees (in programs similar to those for wastewater treatment plant [WWTP] operators) instills pride and promotes healthy competition among the staff.

Competent office and clerical staff members ensure a proper flow of information in and out of the agency and between departments. Computer skills for the maintenance of files and records are required by most agencies. Staff members use computer software extensively and process large quantities of information. Communication skills are important for administrative personnel, who typically are the first contacts for the customer. To keep office morale high, employee performance goals and yearly performance reviews should be established.

Beyond the routine management of the system and personnel, one of the primary functions of administration is to ensure public and employee safety. Federal, regional, state, and local regulations on workplace safety policies can be difficult to track as they come into law, and sometimes they can be difficult to implement into agency policy because of financial, manpower, or political constraints. To keep abreast of changes in workplace requirements, some federal agencies, such as the Occupational Safety and Health Administration, produce regular newsletters with this information as it becomes available. Other avenues of information include various industry associations such as the Water Environment Federation, the American Water Works Association, the American Public Works Association, and the Association of Metropolitan Sewage Agencies. These associations often publish information on laws and workplace requirements that directly affect their members. It is the responsibility of administration to keep all employees informed of changes in workplace safety and provide the necessary training to bring the agency into compliance (see Chapter 1 for additional information on safety and Chapter 7 for additional information on training).

One of the most frequently overlooked aspects of administration is employee development. It is the responsibility of administration to identify and train employees who show potential for advancement into field or administrative management positions. Too often an agency is forced to seek replacement personnel outside the department when a key position is vacated because they do not have a ready replacement trained. Employees are an investment—the longer they remain with the agency, the higher the investment in the employee. An effective employee development plan may save money in the long term by saving the invested dollars in employees who exhibit talents that can be useful in a position beyond their current posting. Among the benefits of employee development are elevated level of morale, better cooperation between personnel and departments, higher degree of loyalty to the agency, and the incentive for personal growth for individuals.

Employee development programs typically have dual purposes that provide global education advances among all employees and identify employees with potential beyond their current levels. Many agencies use identified personnel in management focus groups, such as an internal safety committee, to provide leadership and assist in the development of workable programs or policies for the workforce. Not only does this help administration develop personnel, it fosters better relations within the agency and advances a collective team philosophy.

An effective administration develops a good working relationship with local organizations such as environmental regulatory agencies; health departments; state natural resources departments; police and fire departments; and gas, electricity, water, and telephone utilities. Such partnerships facilitate the exchange of personnel and equipment; mutual support during emergencies and abnormal situations; and the sharing of new procedures, ideas, and techniques (see Chapter 5).

MAINTENANCE STAFF. Proper scheduling and cross training help maintenance crews handle anticipated workloads. Employee skill levels vary with experience, education, and attitude (WPCF, 1987). An established list of priorities and the ability to transfer personnel among different work crews contribute to success. Cross-trained personnel help reduce potential disruptions from vacations, illness, or inadequate staff levels. Reviews of system maintenance activity, typically at budget preparation time, identify needed changes in staff number, crew assignments, procedures, schedules, or use of overtime.

The type of work (cleaning, inspection, maintenance, and emergency and scheduled repairs) determines the number of workers and skill level of personnel required. The number of crews is determined by the personnel available, the type and amount of work, anticipated emergencies, and the current backlog of repairs or construction.

Outside contractors can help alleviate excessive backlog, perform extensive or complex repairs, or meet specialized maintenance needs (see Chapter 6). Costs, which can be substantial, are carefully considered before this alternative is chosen. If the backlog is the result of poor preventive maintenance policy or planning, the agency can address the problem and save money. If the project is beyond the capabilities or experience of the agency, or will take personnel away from regular duties for too long, outside contracting may be a viable alternative.

Collection systems field operations can be broken down into four basic functions: inspection, routine cleaning, emergency operation, and repairs. These operations typically are performed by separate crews in larger agencies, with smaller agencies using the same personnel for these duties. Agencies responsible for pumping stations typically have an additional crew for these stations (see Chapter 4).

Smaller systems typically have fewer, less specialized operators. Costs of such items as large cleaning equipment or video cameras may make it more economical for small agencies to contract the work out or share capital costs with neighboring utilities.

Maintenance staff and field management staff typically are charged with carrying out the policies and procedures established by administration management to field personnel. This includes some training responsibilities involving workplace safety awareness, equipment operation procedures, job duties and responsibilities, information coordination from field operations, scheduling of activities and personnel, and employee evaluation if in supervisory positions.

Inspection Crews. An inspection crew is responsible for system surveillance and documentation. In some cases, the inspection crew is responsible for sealing or grouting work as needed. Small agencies may not have the staff or resources to provide a separate inspection crew, and may instead use maintenance personnel to note easily seen problems and contract inspection services for specific areas. In agencies that combine surveillance and general maintenance, one person is assigned the responsibility for the inspection process and equipment.

In larger agencies, depending on the type of procedures required, the inspection crew may have a minimum of two to three members. Local safety codes may dictate the required number of personnel for confined space entry. Local occupational safety and health agency requirements are therefore evaluated. System surveillance using flow measurements, dye tests, video inspections, and smoke testing is done at regular intervals.

Visual inspections include observing surface conditions along sewer pipe alignments, inspecting manholes from the ground surface through open manhole covers, and internally inspecting manholes from the surface or by personnel entry. Pipe interiors can be inspected visually by lamping from manholes, the use of closed-circuit television, or by personnel entry in large pipes. Surface inspections should document settlement above pipe alignments, buried manholes, and exposed piping in stream corridors.

Surface manhole inspections are performed during routine maintenance operations and during emergency situations. Routine surface manhole inspections are performed to detect the following

- Manhole structural problems,
- Groundwater infiltration,
- Surface inflow through manhole covers or below manhole frames,
- Collection systems cleaning needs, and
- Acts of vandalism.

Manhole surface inspections are also used during emergencies such as system backups, overflows, and odor complaints. Surface observations are valuable for pinpointing the limits of collection systems maintenance difficulties and tracking illegal discharges and odors.

Internal manhole inspections are performed to obtain information that may be difficult to determine from surface inspections. This information includes

- Assessing manhole invert and channel conditions,
- Identifying manhole infiltration and exfiltration points,
- Determining structural condition of connected pipes,
- Identifying hydraulic bottlenecks because of inappropriate pipe elevations or channel configurations,
- Determining effectiveness of collection systems cleaning,
- Measuring pipe diameters and wall thickness,
- Determining the severity of deterioration resulting from chemical and acid contact,
- Measuring pH of manhole and pipe wall materials,
- Recording instantaneous flows, and
- Identifying and quantifying extraneous flows from pipes.

Internal inspections of every pipe connected to every manhole in a typical separated collection system covers approximately 20% of the system. Internal television inspection is used to completely inspect pipes between manholes. Field inspection logs and video tape recordings are used to document pipe conditions (see appendix, Defect Classifications Guide). Figure 3.2 illustrates a typical inspection report that can be handwritten or computer generated. These inspection records can be reviewed and entered into a master data program to determine maintenance or rehabilitation needs and subsequent budgeting of recommended procedures.

The inspection crew typically has a vehicle that will carry or tow all necessary equipment. Examples of this equipment are

- Video inspection camera with recorder;
- Flow and velocity meters, weir kit, and sampling devices;
- Safety harness and other safety equipment;
- Fresh-air blower and smoke test blower with smoke bombs;
- Rulers, surveyors rod, "rolatape," sewer plugs, and air compressor or tank;

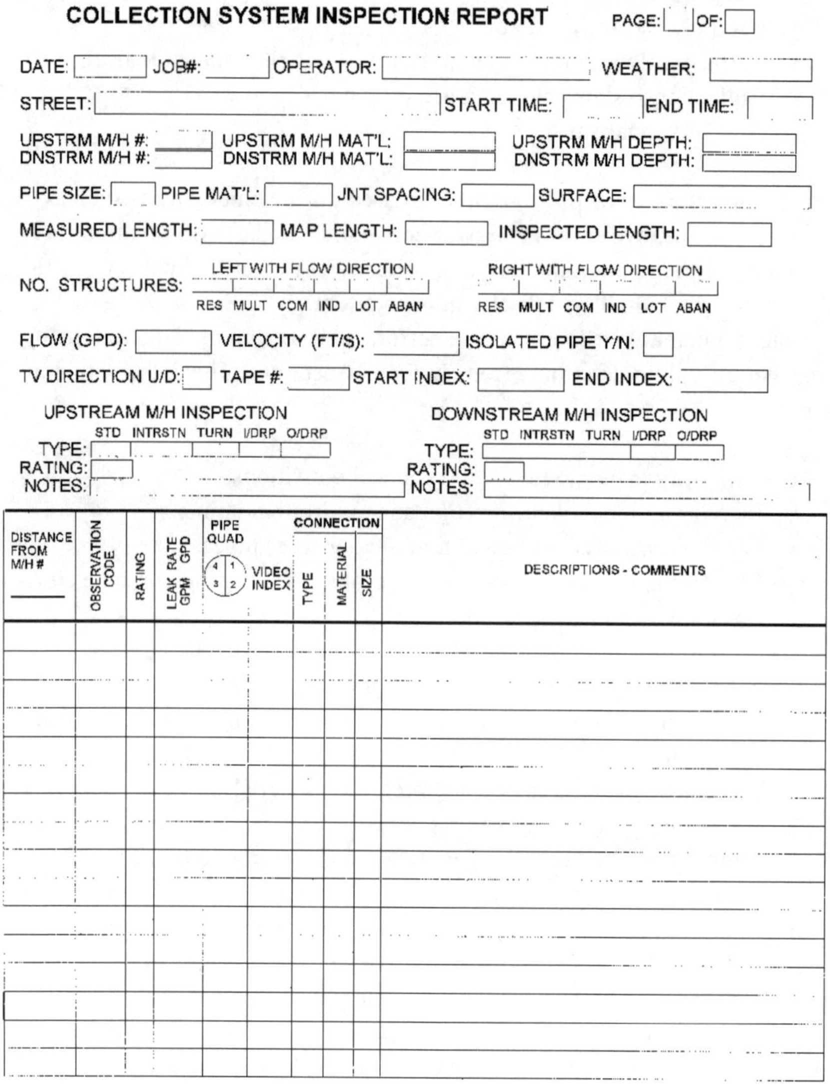

COLLECTION SYSTEM INSPECTION REPORT PAGE: OF:

DATE: JOB#: OPERATOR: WEATHER:

STREET: START TIME: END TIME:

UPSTRM M/H #: UPSTRM M/H MAT'L: UPSTRM M/H DEPTH:
DNSTRM M/H #: DNSTRM M/H MAT'L: DNSTRM M/H DEPTH:

PIPE SIZE: PIPE MAT'L: JNT SPACING: SURFACE:

MEASURED LENGTH: MAP LENGTH: INSPECTED LENGTH:

NO. STRUCTURES: LEFT WITH FLOW DIRECTION — RES MULT COM IND LOT ABAN; RIGHT WITH FLOW DIRECTION — RES MULT COM IND LOT ABAN

FLOW (GPD): VELOCITY (FT/S): ISOLATED PIPE Y/N:

TV DIRECTION U/D: TAPE #: START INDEX: END INDEX:

UPSTREAM M/H INSPECTION — STD INTRSTN TURN I/DRP O/DRP
TYPE:
RATING:
NOTES:

DOWNSTREAM M/H INSPECTION — STD INTRSTN TURN I/DRP O/DRP
TYPE:
RATING:
NOTES:

DISTANCE FROM M/H #	OBSERVATION CODE	RATING	LEAK RATE GPM GPD	PIPE QUAD	VIDEO INDEX	CONNECTION TYPE	CONNECTION MATERIAL	CONNECTION SIZE	DESCRIPTIONS - COMMENTS

Figure 3.2 Typical television inspection report.

- High-intensity lights for manhole inspection and night work site operations;
- Metal detectors, and locating equipment for pinpointing repairs and digs;
- Dye testing powder (various colors);
- Spool of sectional rods with a tool for clearing stoppages;
- Spare manhole cover and cover seals;
- Small aluminum ladder and canvas buckets for tools;
- Small pump and hose;

- Assorted hand tools for mechanical repairs; and
- Mirrors and assorted lights.

Other basic items are communication radios, personal safety attire, and road cones. This equipment is examined frequently and repaired or replaced as necessary.

The inspection crew primarily is responsible for locating and noting problems that need the attention of either the general maintenance or repair crews. Visual inspection with a preventive maintenance program facilitates the detection of minor problems in the system before they become major. By planning downtime in the system, the crews can make repairs under controlled conditions for less cost than repairs under emergency conditions.

Regular Maintenance Crews. Regular maintenance crews clean and remove obstructions from the sewer lines, provide regular checks and cleaning of known problem areas, and carry out minor repairs and special programs such as chemical dosing or daily system point checks. Small agencies may contract some specialized tasks to specialized contractors.

All members of the crew should be cross trained, with the ability to perform multiple cleaning tasks and operate several pieces of equipment such as high-velocity cleaning machines, rodding machines, vacuum removal machines, and bucket machines. Some regular maintenance crew members should also be trained in repair techniques and closed-circuit inspection equipment use.

In some instances, especially for small agencies, it may be not be financially practical to duplicate equipment, backhoes or dump trucks for example, used by another agency within the same government body, such as the street department. Equipment sharing is not considered a good solution unless a single agency is responsible for both collection systems maintenance and other functions, for instance a public works department handling both street maintenance and collection systems. Problems in sharing equipment between separate departments can develop when both departments need equipment at the same time, often creating friction. Sharing equipment provides short-term financial benefits and, if managed properly, can help coordinate tasks between departments. The question of responsibility may surface, however, when equipment needs repairing or in determining which agency gets priority if each has an emergency at the same time.

Regular cleaning to promote uninterrupted service includes rodding to remove roots and grease; high-velocity flushing to remove deposits in the lines; vacuum and bucket machine operations to remove large amounts of material from trunk and interceptor lines; and chemical or biological applications for grease, roots, and odors.

Some of the equipment used by the maintenance crew are

- High-velocity cleaning machines (truck- or trailer-mounted);

- Rodding machines (truck- or trailer-mounted);
- Vacuum equipment (truck- or trailer-mounted);
- Sewer balls or scooters with winches and stands;
- Bucket machines (trailer-mounted);
- Tripods and harnesses for manhole entry;
- Long-handled forks or rakes;
- Job-site safety equipment (for example, road cones and hard hats);
- Bypass pumps with hose;
- Assorted tools for equipment repair;
- Portable generators with lights;
- Manhole ventilators;
- Atmospheric testing equipment;
- Aluminum ladders, ropes, and canvas buckets; and
- Assorted hand tools such as shovels, picks, and brooms.

It is important to train crews thoroughly in the operation of high-velocity cleaning equipment, rodding machines, and specialized equipment such as bucket machines or hand-rodding equipment. A detailed discussion of cleaning this equipment is given later in this chapter. Safety dictates attention to proper setup, operation, and maintenance. New employees not already equipped with broad maintenance experience can be trained in specific uses of new power equipment. It may be efficient to have specific crews for the high-velocity cleaner or rodding machines because practice and technique are important to their safe operation.

Each piece of equipment is examined regularly and replaced when it becomes more expensive to operate and maintain than a newer piece. Large machines have a replacement schedule that anticipates the projected lifespan of the equipment. Planning or budgeting based on the anticipated lifespan of equipment helps avoid unexpected equipment failure and can allow the cost of purchases to be spread out during a period of years. See Chapter 8 for more details on financial planning.

Activities of the maintenance crew are key in the preventive maintenance program. Observations during regular cleaning operations generate records of problem areas so that they can be repaired or serviced. Figure 3.3 illustrates a typical field maintenance record form used with either a paper system or with computer generation. The regular maintenance crew spots problems before they become serious and helps manage problems that arise in the field. Minor collection system repairs, including manhole repairs and small-scale excavations to repair or replace broken pipes, can be made by the regular maintenance crew or by a separate repair crew.

Many agencies use regular maintenance crews for interagency operations such as an industrial waste division or engineering department. A good working relationship between departments can provide solutions to many system problems and can decrease the time needed for enforcement, site studies, and construction or repairs. Routine maintenance crews can provide

FIELD MAINTENANCE RECORD

Date:__________ Street:______________________________ From MH #:______ To MH #:______

Crew leader:______________________________ Time start:______am/pm Time end:______am/pm

Equipment used:______________________________ Materials used:______________________

Weather conditions:__________ Flow volume (GPD):________ Measured:___ Approximate:___

Maintenance work:

Reason for work: __ Scheduled maint. __ Periodic maint. __ Service request __ Emergency
Cleaning footage:__________ Materials removed:______________________________
Comments:____

Evaluation:
Pipe size:___ Condition: __Good __Fair __Poor __ Damaged, describe:______________________

Note defects in manholes:
MH #:______
Barrel condition: __ Good __ Fair __ Poor __ Damaged, describe______________________
Rim/lid condition: __ Good __ Fair __ Poor __ Damaged, describe______________________
Invert condition: __ Good __ Fair __ Poor __ Damaged, describe______________________
Infiltration evident: __ Yes __ No

MH #:______
Barrel condition: __ Good __ Fair __ Poor __ Damaged, describe______________________
Rim/lid condition: __ Good __ Fair __ Poor __ Damaged, describe______________________
Invert condition: __ Good __ Fair __ Poor __ Damaged, describe______________________
Infiltration evident: __ Yes __ No

MH #:______
Barrel condition: __ Good __ Fair __ Poor __ Damaged, describe______________________
Rim/lid condition: __ Good __ Fair __ Poor __ Damaged, describe______________________
Invert condition: __ Good __ Fair __ Poor __ Damaged, describe______________________
Infiltration evident: __ Yes __ No

Flow characteristics:
__ Normal flow __ Sluggish flow __ Flow blocked __ MH surcharged
__ Lateral blocked __ Pipe failure __ Vandalism __ Hydraulic bottleneck
Conditions found:____

Other Comments:____

Figure 3.3 Typical field maintenance report.

information to the varied departments when they encounter mechanical failure potential or industrial discharges. Maintenance crew training should be considered for observations and recording or notification procedures that can be efficiently used for interdepartment communications.

Emergency Crews. Emergency crews are often the most difficult to organize because their tasks are varied and unpredictable. Problems range from a minor complaint to serious flooding, collapsed sewers, explosions from collection systems gases, pumping station failure, or rescue operations. Emergency crews are most effectively led by a person who favorably represents the workers and the agency to the public and has received specialized training in these aspects of the collection system.

In smaller communities, members of the regular maintenance crew who have received special training may constitute the emergency crew. In larger agencies, specialized crews are organized around the types of anticipated emergency calls.

In most instances, emergency response is the direct result of a system user complaint, ranging from basement flooding to vermin activity. Some complaints, such as noisy manhole covers or odors, can be responded to by the regular maintenance crew during the course of their daily activities. Therefore, it is important that both emergency and regular maintenance crews receive training in both problem identification (such as detecting whether an odor is a natural gas leak or hydrogen sulfide) and in public response (such as dealing with a distraught property owner whose land has been flooded).

System complaints should be recorded and tracked, similar to inspection and routine maintenance records. Tracking this information can lead to a more comprehensive maintenance program when paired with other crew records in determining where system repairs or redesign may be needed. Figure 3.4 illustrates a typical complaint response form used with either a paper- or computer-generated system for tracking these problems.

Specially outfitted emergency vehicles prevent delays in locating or loading correct equipment. Examples of emergency equipment are a portable generator with lights and power tools; an air compressor with hose and tools; a rodder or flushing machine; an air blower; bypass pumps with hoses; a backhoe or other excavation equipment; hand tools for digging and mechanical repair; a tripod and harness and other confined space gear; ladders, assorted ropes, and buckets; and communication equipment. Many agencies will have an instant camera or video camera available to document a situation or damage that may result in a claim.

Emergency equipment is carefully and regularly inspected for any sign of disrepair. All equipment is operated periodically if regularly left idle so that it will be kept in top running condition—clean, stocked, and available for immediate use.

Emergencies involving pumping stations typically are handled by the regular pumping station mechanics and electricians. These crews should have appropriate tools and a generator to restore power to a pump in the event of power failure. Pagers, regular and cellular telephones, and radios are used to contact anyone on call.

Special emergency procedures are established to give guidance to those who might suddenly find themselves involved at whatever hour with an emergency. During regular crew meetings, procedures are reviewed and revised if necessary to meet a new situation. These procedures define the circumstances under which other public agencies are notified, how flooding situations are handled, and methods for reconstruction. Untrained personnel are always accompanied by experienced personnel when undertaking emergency operations. See the Stoppages and Obstructions section of this chapter for more information.

COMPLAINT RESPONSE FORM

Date:__________ Time of complaint:________ AM/PM Complaint rec'd by:________________

Street location:____________________ Cross street:____________________

Complaint made by:____________________ Address:________________ Tel: ________

Complaint details:____

Field Response/Evaluation

Time arrive:______ Time depart:______ Foreman:________________ Weather:________

Complainant contacted on arrival: __ Yes __ No Property damage: __ Yes __ No

If yes, describe:____

Cite discharge point: __ Sanitary fixture/s __ Floor drain __ Exterior grounds __ Catch basin __ Other

Check nature of complaint:

__ Sewer stopped	__ HC stopped	__ CB stopped	__ Culvert stopped
__ MH cover off	__ MH cover broken	__ MH cover noisy	__ CB cover off
__ CB cover broken	__ Odor complaint	__ Pump sta. alarm	__ Power off
__ Street flooded	__ Sewage leak	__ Basement flooded	__ Debris on street
__ Street depression	__ Insects	__ Vermin	__ Lost article

Observation/response: (check appropriate items) Upstrm MH #:_____ Dwnstrm MH #:____

__ Normal flow	__ Sluggish flow	__ Flow blocked	__ MH surcharged
__ Lateral blocked	__ Pipe failure	__ Vandalism	__ Hydraulic bottleneck

Conditions found:____

Maintenance work performed:

Affected from manhole #:_____ To manhole #:_____ # feet:_____ Bypass pumping used: __ Yes __ No

Line cleaned: __ Yes __ No If yes, method used:________________

Material removed:________________________________

Problem relieved: __ Yes __ No

Actions taken:____

Follow up recommended:

__ Owners problem	__ TV inspection needed	__ Rodding needed	__ Flushing needed
__ Repairs needed	__ Inflow disconnect needed	__ Pest control needed	__ Unknown problem

Comments:____

Figure 3.4 Typical service request/complaint response form.

Repair Crews. Repair functions and maintenance functions for many small agencies can be interrelated using the same personnel for a variety of field repairs, or if the workload is significant, separate crews can be established. There are a number of conditions under which an agency may want to take corrective action. The collection systems manager needs to consider several questions before starting any project.

- Is the project beyond the ability of available personnel and equipment?
- If so, will the experience gained by doing it save future costs?
- Is there enough time to do the job properly?
- What is the estimated cost of the agency performing the repair versus the cost of contracting?

- If the project is attempted, what normal operations may suffer by reassigning personnel to do it?

The size of the agency may determine which repairs are considered minor and which major.

For many communities, annual unit price or a time and materials contract for underground repair work may be a more financially attractive alternative to obtaining heavy equipment and training agency personnel. Larger agencies may need one or more full-time repair crews responsible for such things as replacing collapsed lines, repairing masonry in manholes and large-diameter pipe, raising and lowering manhole covers, laying new sewers, and installing cleanouts and backwater valves. Some repair personnel can be assigned to the emergency crew for excavation work and handling heavy equipment and materials.

Beyond equipment needs for the repair crew, materials to affect repairs should be on hand for emergency situations to avoid potential delays in performing emergency repairs. The maintenance yard should have a small inventory of bedding gravel, sand, and concrete and mortar supplies to allow a small excavation repair to be done after hours or during weekends without the need to purchase the materials before starting. A small inventory of pipes and pipe fittings in average sizes should be available for emergency use. A typical inventory of pipes would be several metres of 8-, 10-, 12-, and 15-in. sections. Depending on the system, these pipe sections may need to be of varied materials such as polyvinyl chloride (PVC), vitrified clay, or concrete. Some agencies will use PVC for most repairs, using adjustable rubber clamps to join into older pipes of different materials, while other agencies will stock pipe in all types typically found in the system. In addition to pipe sections, a small inventory of elbows, connectors, and splice fittings of various sizes should be on hand. Some small lengths of pipe with fittings for typical residential and commercial service connections should be available in the event they are needed during unscheduled repairs. Historical data on past repairs may offer insight as to the most frequent line repairs, allowing some level of prediction on what will be needed during a given time.

As a rule, repair crews are led by someone experienced in field repairs, and repair work is not started without experienced field help. The staffing of the repair crew varies with the size of the agency. A typical crew consists of a crew leader, backhoe operator, maintenance person I, maintenance person II, and a laborer.

Repair crews frequently are tasked with working with inspection crews to provide service work in manholes and internal pipe repairs. This can include manhole grouting, point repairs, grouting pipe joints, exploratory excavating, and pipe relining.

Backhoes, air compressors, shoring or trench boxes, front-end loaders, dump trucks, generators and lights, tampers, pumps, assorted hand tools, a transit and grade pole, and a utility truck with assorted mechanical tools are typical repair crew equipment.

Unless the agency is large and has the financial resources to purchase heavy equipment, tasks requiring large shovels or specialized excavation and other equipment may necessitate help from another utility or outside contractor. For some municipalities, much of the smaller equipment is also needed by other municipal agencies. Because duplication is costly, sharing equipment between departments may be financially attractive as a short-term solution if the agency is willing to be flexible in scheduling excavation work.

Equipment should be well maintained and kept in safe operating condition. Replacement policy typically dictates that equipment be retired at the end of the projected lifespan or when the operating cost greatly exceeds that of a newer piece of equipment. Thus, the agency is not forced to operate old, out-of-date, and potentially dangerous equipment. Many municipalities operate on a 5- to 8-year capital plan that anticipates and budgets for equipment needs 5 or more years in advance of the purchase date.

The equipment required to operate collection systems can be expensive, and it is recommended that the agency examine the options in equipment acquisition. Specialized or infrequently used equipment can sometimes be rented or borrowed. Alternatively, the agency can contract out for the service. Regularly used equipment can be leased or purchased outright. There are advantages to the outright purchase of equipment with a long lifespan; however, the equipment will have to be kept in good condition for many years while funds are raised for replacement. Leasing is attractive because it allows equipment to be obtained at a lower initial cost. Leasing terms can be flexible enough to allow the agency to avoid operating out-of-date equipment should there be advances in technology (see Chapter 8 for more information).

When engaging in repair projects that will affect system service or local traffic in a small area, many agencies provide local notification through door hangers to explain what is happening, what to expect, and how the project will benefit the community. Some projects may affect a large area or main traffic routes, in which case a statement may be issued to the local press to advise the public of the planned activity. Though somewhat time consuming, this public relations effort can have excellent results in reducing complaints from the affected area and enhancing the profile of the agency among its customers.

Construction sites, especially in highly traveled areas, need to be secured during working hours. The use of barricades, cones, or traffic control may be necessary to prevent traffic accidents or injury to field personnel. Additionally, if the job site must be vacated for any period of time, efforts to secure the site should be part of the operating procedure. Securing excavations so that curious pedestrians cannot fall into open pits or trip over debris should be a primary task. If the excavation cannot be sealed with metal plates, proper traffic and pedestrian barricades should be erected to prevent motorists or pedestrians from entering the closed site. Equipment should be removed from the site if possible or secured from theft, vandals, or inadvertent operation during nonworking hours.

FACILITIES AND EQUIPMENT

A properly planned and supported equipment yard is essential to collection systems operation. It is the location from which equipment, materials, and personnel are dispatched and the location where operations records are kept. Ideally, these facilities are self-sufficient and independent, except where a central repair yard or heavy-duty repair shop is available.

In smaller municipalities, collection systems maintenance equipment and personnel typically share one yard with other municipal operations, such as water and street departments. In this case, it is important that a portion of the yard be designated for collection systems maintenance activities alone. Specialized tools can be segregated or color coded so that they are not used on water supply systems. This reduces confusion regarding equipment use and repair. All essential tools and portable equipment should be kept in a secure location, with keys issued to supervisory personnel.

The area should be planned carefully; often the yard is small, and the opportunity for confusion increases as the space diminishes. If facilities are properly placed, interference between departments is reduced. The layout and location of the yard are reviewed based on current use and traffic conditions and current and future land values.

Satellite yards to house specialized collection systems equipment can reduce response time on weekends or holidays. Proper response to collection systems complaints will improve public relations; the primary goal is to restore service promptly in the event of interruptions. In large cities, considerable time can be lost because of traffic delays or distances between the yard and the job site. Agencies in growing metropolitan areas typically prepare a long-range program and divide the area into districts, each with a central yard.

Yard design (Figure 3.5) should provide the proper housing of equipment and materials affected by inclement weather. Special storage space can be allotted for concrete, bricks, and other small materials. Raised platforms inside and outside accommodate the on- and off-loading of daily equipment, materials, and vendor deliveries. It is recommended that standard or special platforms match the truck bed level. A platform-mounted, hand-operated hoist is useful for heavy lifting.

Pumps, compressors, mixers, generators, bypass piping, and other items are best stored separately so that available equipment is visible at a glance. Emergency equipment used at irregular hours (lights, bypass pumps, and so

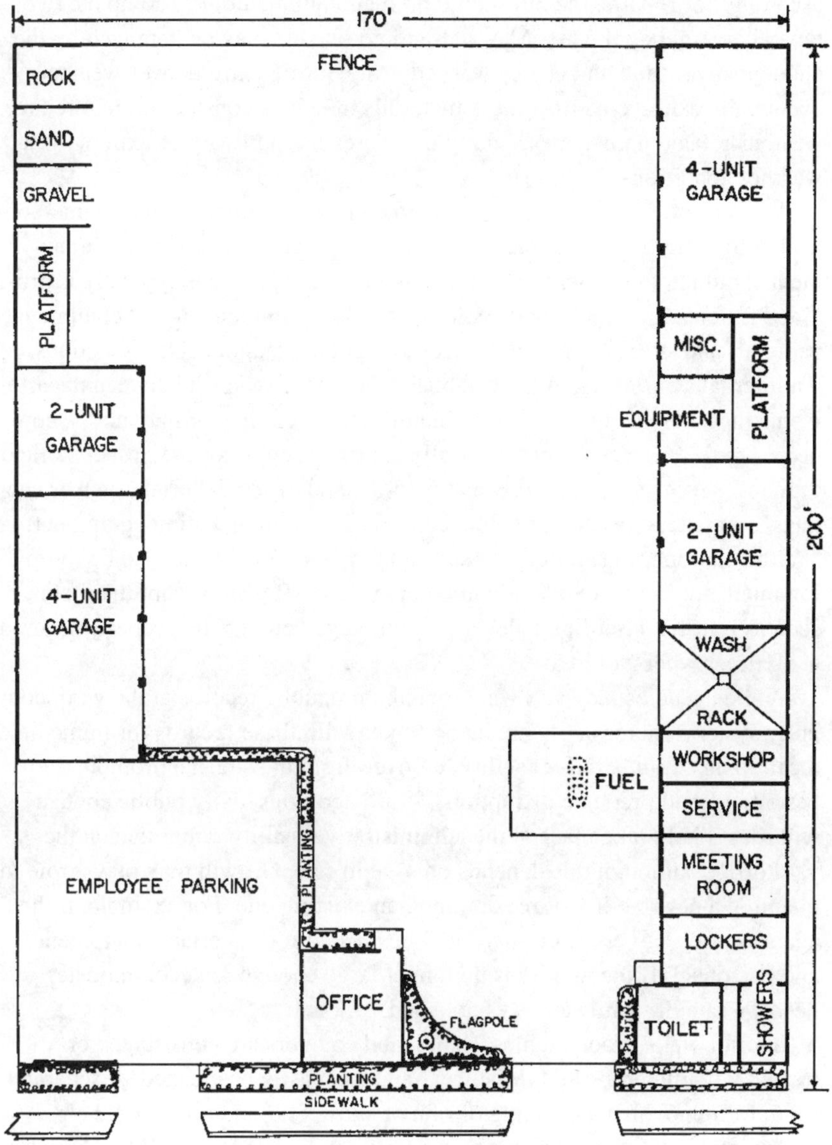

Figure 3.5 Typical sanitary maintenance yard, showing garages, office, personnel locker room, and vehicle fueling facility.

on) typically are kept away from regularly used tools so that they can be found quickly when needed.

Storage bins for sand, gravel, dirt, and asphalt should be wide enough to fully accommodate the bucket of a front-end loader or backhoe so that the material may be quickly handled and moved. The bins typically are deep enough to hold at least a 1-month supply of materials if used regularly; if used only for repairs, the bin should hold enough material to complete two typical emergency repairs. Materials stored outside may be damaged by the elements over time and can be washed away during extreme wet-weather events. Providing cover for these materials may be a consideration to reduce weather-related damage, especially if the area has a history of extreme wet-weather conditions.

A clean and orderly yard can strengthen morale and efficiency. Time-controlled lighting for safety, security, or night operations is essential. Equipment should not be stored dirty or in disrepair. A wash rack typically is provided for cleaning trucks and equipment. It is recommended that equipment, vehicles, and storage areas be subject to regular unannounced inspections. An internal award system can establish a competitive spirit between departments that share a common yard. The office, which may or may not be connected to the other buildings, typically is large enough for the proper storage of maps; survey charts; reports and records; and office materials such as cabinets, computers, printers, reference materials, communication equipment, and the appropriate furniture for staff and visitors. There can also be room for small, specialized tools or computer systems. It is important that items such as transits, laser lights, levels, and footage counters be properly secured at all times when not in use.

While agencies may still want to maintain public records at the yard, computer systems increasingly are being linked with these records for immediate recall to an administrative facility removed from the site. To promote the best image and preclude disruption of yard activities, daily public contact will most likely take place at the administrative facility rather than at the yard office, although this depends on system size. Growth may play a role in planning a new facility or reexamining an existing one. For example, if the agency grows larger, more space will be needed for materials, equipment, and personnel. If the facility is no longer large enough to accommodate needs, a satellite yard can be established.

Interior walls, floors, tables, chairs, and counters are constructed or treated with materials that resist wear and can easily be cleaned. A separate room for removal and cleaning of rain or snow gear and boots will help lengthen the life of this equipment and keep the rest of the facility clean. It is important to keep in mind that the administrative offices, whether in a central or satellite facility, are open to the public. A clean and orderly appearance can be the best public relations image an agency can present.

Central repair maintenance shops vary in size and scope. Some agencies maintain the sanitary collection system only, while others are responsible for storm sewers, pumping stations, treatment plants, or street repairs. In smaller communities where street, water, and collection systems departments are consolidated in public works, it may be more economical to have a central repair shop. At the common yard, all equipment maintenance would be performed and all departments would operate. A utility truck can be set up with an assortment of tools; parts; and welding, cutting, and hoisting equipment so that the mechanic can go to the problem when necessary.

BASIC MOBILE EQUIPMENT. A current trend is to outfit trucks to fit the needs of the project or meet the demands of the specific job. Operations are more efficient with the proper types of tools and trucks. The same is true for the maintenance of pumping stations at which electrical and mechanical work is performed.

Many agencies use open 1-t (1-ton) or smaller pickup trucks. These vehicles are convenient and durable but provide little protection from inclement weather or sun. Small or expensive items can be lost from the vehicle if not locked in a storage box. Utility truck bodies have lockable side compartments for equipment and tools, with an open payload area. A canopy can be installed over the body to protect equipment and materials from the weather. Specific adaptation of this type of vehicle increases efficiency and presents a pleasing appearance.

Ideal construction trucks are rugged, compact, and equipped to carry the necessary equipment for routine and emergency repairs. Truck size and design are determined by the agency's budget and the task, though trucks can be set up for dual duty to save expense. For example, with the addition of a generator and lights, the pumping station repair truck can also be used for emergencies occurring after normal working hours.

Trucks for masonry and construction have separate compartments to prevent mixing of materials. These trucks typically have an area for debris and spoil from the job, if practical. If not, a truck is designated for hauling debris. Trucks for regular cleaning can store necessary cleaning tools, preferably in lockable storage compartments. Long-handled tools such as sewer rakes and shovels can be stored on the vehicle. A trailer hitch designed to support at least 4 500 kg (10 000 lb) can be mounted on the truck for towing.

Vehicles should be equipped with two-way radios so that regular contact can be made with the office and advice, instructions, or emergencies can be communicated. Trucks can also be equipped with strobes or beacon lights for high visibility and traffic control applications. These vehicles typically are painted a bright or reflective color and have reflective stripes for low-visibility conditions. In heavy traffic areas, a lighted directional arrow to direct traffic is useful.

SPECIAL MOBILE EQUIPMENT. As the sanitary collection system expands and ages, specialized maintenance equipment becomes necessary. This equipment can be costly, and it may be more cost effective to use modified standard equipment. There are two varieties of pipeline cleaning equipment—mechanical and hydraulic. Mechanical cleaning involves the use of a tool (such as rodders, bucket machines, and hand rods) to clean the pipe. Hydraulic cleaning (balling, scooters, and high-velocity cleaners) uses water to move material down the pipe. The rodder and high-velocity cleaner are probably the two most common pieces of collection systems cleaning equipment.

In larger systems, backhoes, trenchers, loaders, and dump-bodied trucks are desirable for underground repairs. Agencies that do not use this equipment frequently may find it costly to purchase and maintain, and outside contracting may be preferable. Mechanical equipment can reduce the costs of excavation, excess dirt removal, and backfilling. The use of power equipment can reduce the cost of time and labor. The number and types of jobs determine whether the machines are purchased or rented.

In systems with extensive off-road easements or access roads to maintain, all-terrain vehicles can be used to enable routine inspection of remote areas not easily reached or traveled by conventional vehicles. Trailers can be outfitted for use to allow transportation of tools and materials to remote access areas.

Acquiring any of this equipment requires planning and recordkeeping for establishing a capital purchase plan. This plan includes how and when to make the acquisition. Planning equipment purchases can preclude the need to replace failing equipment before money is available or research into new equipment can be conducted.

PORTABLE EQUIPMENT. It is important that collection systems maintenance be geared to a 24-hour-per-day, year-round operation. Portable equipment must be moved quickly and operated under extreme climatic conditions. If not used regularly, this equipment should be operated on a routine basis to ensure readiness. Equipment requirements are determined by system size and type of task. Air compressors with a minimum capacity of 70 L/s (150 cfm) are basic equipment because they can power numerous pneumatic tools.

Because repairs on broken lines must be completed after they are initiated, generators are essential to permit night work. Safety considerations require that adequate lighting be available at the job site. Any electric tool can be used in the field. Pumps can evacuate water from collection points such as trenches, manholes, wet wells, sumps, or basements. Gasoline- or diesel-powered pumps with a discharge size of 5 to 20 cm (2 to 8 in.) can be used depending on the size and volume of sewers to be serviced. Self-priming and nonclogging pumps are most effective. Small pneumatic pumps are good for high-lift, low-volume applications such as pumping out basements or dewatering trenches. Larger pumps are used for bypass pumping or for draining large areas.

Confined space entry is dangerous without ventilation and atmospheric testing. Air blowers decrease the dangers of explosive or other hazardous gases and provide oxygen in confined spaces. Air blowers, which are gasoline or electric powered, come in differing outputs. The required output is determined by the structures the agency has to ventilate. Blowers designed to ventilate at a rate of approximately 3 500 L/s (7 500 cfm) are effective at ventilating a medium-sized sanitary sewer. Small portable units that can be carried on the truck typically supply ventilation at rates of between 350 and 570 L/s (750 to 1 200 cfm) and are adequate to ventilate most manholes.

In the event of a broken pipe or other condition causing discharge, emergency chlorinators are important for disinfecting wastewater that flows directly to streams and ponds. Prolonged discharge of raw wastewater to receiving waters is at minimum a nuisance and at worst a health threat to people and wildlife, though chemical disinfection can also present environmental problems if not done properly. Disinfection at the discharge site should be provided by qualified, trained personnel and follow the guidelines of the state environmental agency.

Other examples of specialized portable equipment are asphalt saws, shoring equipment, concrete mixers, gasoline-operated power saws, and chemical spray units (for insect and rodent control). Table 3.1 lists equipment typically required for small agencies (approximately 260 km [160 miles] of sanitary sewer and responsibility for storm drains).

LOGISTICS. Logistics are important to efficient maintenance, especially in a large system or where an extensive range of equipment and tools is carried in the inventory. If the necessary replacement parts or materials are not on hand or cannot be located, a simple repair job could, in some cases, escalate into an embarrassing emergency. The situation is aggravated if there is an unanticipated long lead time for a part that has been ordered for an equipment breakdown. Inventory control can be used to ensure timely acquisition and ample stock of parts. Many agencies use computerized inventory systems that provide automatic notification for ordering materials or parts when a minimum inventory level is reached.

Logistics play the lead role when repair projects are undertaken. Logistical project support includes coordination of materials delivery to a project site so work can continue without scheduling delays. This is probably the most difficult of responsibilities to effectively manage, especially if the project is in a high-profile area or will interfere with traffic or commerce. If materials are needed, bids need to be coordinated well in advance of the project, with delivery requirements and monitoring; contingency plans should be addressed to cover unexpected downtime if equipment fails; and public relations materials need to be assembled for the news media. Failure to plan for these items can sometimes lead to problems if the project is halted because of lack of materials, equipment problems, or lack of understanding by the public affected by the project itself.

The maintenance history of equipment, life expectancy of system components, and condition of the system are considered when setting up the program and ordering stock. It is important to know the sources of parts and materials and the time necessary to secure delivery. Most agencies can handle this task with the help of a computer. Computer costs are now such that even small agencies can use them. Inventory control software is available using

Table 3.1 Sewer maintenance equipment for small municipalities.

Item	Purpose	Remarks
Truck: 1 Mg (1 ton) with enclosed body and open center space	General purpose	Stock item: mount and carry equipment, materials, personnel, tow trailers
Small dump truck	Repair/installation construction and hauling capabilities	Optional: cartage for in-house construction projects basis
Backhoe	Excavation for repairs/installation	Optional: allows in-house excavation
Small trailer	Carry tools, materials, or construction debris	Stock item: transport of tools, heavy or bulky equipment
Air compressor, trailer mounted	Small construction/repair projects, facility maintenance	Considered stock item: allows operation of air-activated tools
Portable winch, hand or powered	Stringing lines, controlling sewer balls	Optional: considered stock item in most instances if televising or balling are used
Portable generator	Power source for lighting and electrically powered equipment	Stock item: provides emergency power in remote locations
Portable lighting	Provides lighting for work sites	Stock item: used during night operations; use explosionproof if used in underground applications
Sewer rodding machine, trailer or truck mounted	Clear stoppages, removal of roots and grease	Stock/optional: considered stock item for many systems; can be used in conjunction with flusher if available
High-velocity flusher, trailer or truck mounted	Routine maintenance; removal of sediments, grease, and stoppages	Stock/optional: considered stock item for most systems as a P/M tool; as with rodder, can be used exclusively

(Continued)

Table 3.1 Sewer maintenance equipment for small municipalities (continued).

Vacuum equipment, truck or trailer mounted	Debris/solids removal from manholes or facility maintenance	Optional: allows removal of sediments in lieu of surface scoops or confined space entry
Hand tools; spades, manhole hooks, repair tools, ladders	General purpose	Stock items: used in typical daily field activities
Gas testing equipment	Testing of confined space for toxic, explosive conditions	Stock item: used to test confined spaces before entry, and for monitoring air quality
Self-contained breathing apparatus	For use in hazardous environments	Stock item: allows hazardous environment entry or rescue
Air blower	For use during confined space work	Stock item: used in manhole ventilation
Traffic safety equipment	For motorist/crew protection in exposed work sites	Stock item: barricades and cones will depend on local conditions and traffic
Personnel safety equipment; protective clothing, gloves, hard hats, eye protection, confined space entry winches	For daily use in operations in meeting mandated safety regulations and Occupational Safety and Health Administration requirements	Stock items: provide worker protection for self and others in daily field work

bar coding, or outside consultants can be used to construct an adequate program. After all of the information from paper records is loaded into the computer, the system can signal when repair parts should be ordered for specific machines and tell the manager what is on the shelf at any given time.

PREVENTIVE MAINTENANCE

The primary goal of collection systems operation is to transport wastewater uninterrupted from the source to a treatment facility, keeping the wastewater in the system while keeping groundwater out. A collection system can represent the single largest investment in a community and should be protected and cared for accordingly. Information recall and management methods will play a critical role in reaching a high level of system performance.

When equipment and manpower needs have been determined, the next step is to plan how these resources will be used. There are three different types of management philosophies used in collection systems operations—crisis, preventive, and predictive management. With crisis management, agencies perform maintenance primarily in response to emergencies. Because crisis management is characterized by little if any planned maintenance, problems tend to grow. Emergency calls build in number as the causes of flow interruption increase, eventually allowing focus only on the emergencies and not the causes. Crisis management is more costly because expanded overtime and use of outside contractors can be required, sometimes just for routine work, when regular crews are engaged in emergency services or repairs.

Preventive maintenance management includes planned and scheduled line maintenance or repairs based on past history, and includes the systematic inspection and cleaning of the entire system. Preventive maintenance can reduce operation costs during both the short and the long term because it is far less expensive to maintain system appurtenances than to repair or replace them. Additionally, with increased attention to the system, the incidence of backups can be greatly reduced and result in savings in labor and cleanup costs, not to mention potential liability costs. In using a scheduled approach to system maintenance and repairs, budgeting can become more accurate, with allocations being directed to specific operations or programs designed to allow maximum life of the system components.

Preventive management is a discipline that is pursued on a daily basis. Data from routine maintenance, inspections, flow measurements, and complaints are all entered into the database system. With the many thousands of variables possible with these data, paper-based systems are being phased out in lieu of computer applications.

Predictive management is the attempt to manage a system based on historical information and how it ages. All system appurtenances such as pipes, pumps, and structures will deteriorate over time. Predictive management is an important feature of preventive maintenance in identifying problem areas, and predicts what the priority of maintenance, repairs, or replacement will be during a given length of time. Predictive management uses all the facets of preventive maintenance and can be used for both long-range replacement or repairs and for establishing routine maintenance work orders for areas with known histories. Again, recordkeeping is the basis of an effective predictive management program, without which even the best guesswork will not produce the desired long-term results. For agencies with limited manpower, equipment, or financial resources, predictive management can be an effective program to keep ahead of problems that can cause major repairs or flow interruptions, and spreads the costs of remedial work over time. As with preventive management, funding allocations are needed over time to allow predicted work to occur on a planned basis.

A preventive maintenance program is one of the best ways to keep the system in good repair and operation. A good preventive maintenance program depends on properly trained personnel (see Chapter 7). State certification programs, correspondence courses, technical schools, and in-house training programs are all sources of essential training to ensure the best results.

SCHEDULING. Scheduling permits regular system service and prompt problem location and creates a working plan for the system as it grows and ages. Scheduling, a basic ingredient in the preventive maintenance program, requires information.

External information from system users (the public) is particularly important to scheduling because customers are often the first to spot a problem, such as a backup, and are essential to early detection. Ideally, a complaint is addressed and corrected as soon as possible, with the resulting data added to the collection systems maintenance program. This can include the address or location of the problem, time of day, type of problem, and weather conditions. This information is recorded on a service request form (Figure 3.4) on either a paper- or computer-generated format. These field records can be entered into a database for tracking by date, location, problem cause, and name and address, and used to generate work orders and follow-up calls if needed.

Information recall becomes important when assessing data for preventive maintenance. Paper systems typically track by form number, location, or problem cause. Every agency has different criteria for these forms, typically developing their own in-house paper form for fieldwork. Computer applications often can be customized to reflect needed data entry for specific systems and can be recorded using paper records from the field or direct input from field computers (laptops).

Information from internal sources such as maps, prior reports, or field experience is also important in scheduling, as are other factors such as pipe size, construction, flow volume, service area, and past history. As-built drawings or construction plans are good sources of information about, for example, grade, soil composition, or depth. These references can also indicate bottlenecks and other hydraulic problems not visible from the surface. Geographical information systems (GIS) are becoming a widely used tool in managing collection systems problems and service history data used for scheduling. If the agency is planning to computerize, or is computerized, GIS should be considered for highly accurate mapping and detail of the system.

Field information is probably the most accurate type of information. Field crews should have standard forms (Figure 3.3) to report activities, damage discovered, and conclusions. The importance of training cannot be understated for accurately recording actionable repairs and maintenance work. This information is subject to the interpretation of the person recording the information. Only through training will all reports follow a specific format, with global reference codes used that can be understood by management, engineering (internal or contracted), and field personnel. After the agency format is determined, field personnel can be trained to use data codes to help increase accuracy and speed in processing the information.

Organizing the workforce on a daily basis can be the most difficult job in collection systems operation. Collection systems run more smoothly when the frequency, manpower, and equipment needs for O&M and the facilities owned and operated by the agency have been cataloged. With these lists and the information from the field reports, the collection systems manager can start to build a solid preventive maintenance schedule. It is important for the system supervisor to have a good working knowledge of the equipment involved to determine the time and manpower required for a given task. Basic assigned duties should include daily checks of known trouble spots in the system. Along with these checks are other routine system inspections that need to be done on a frequent basis, such as siphons, bottlenecks, and pumping stations (if applicable). All these areas need to have scheduled inspections, typically based on past historical information.

Personnel typically are assigned to tasks according to priority. For example, line cleaning in the commercial food district should take priority over clearing brush from easements and rights-of-way, and raising manhole covers for new road surfacing would be a higher priority than routine camera inspection. Personnel scheduling fieldwork need to consider time needed for safe travel, traffic, setting up and taking down equipment, and weather conditions. Workers are needed in the field to meet applicable requirements for confined space entry (see Chapter 1). Confined space entry typically requires gas-detection meters, air blower, tripod and winch, safety harness, and personal safety equipment or apparel.

The use of computer applications in collection systems management has grown exponentially during the last decade. Computer systems can be relatively inexpensive and, with the proper software application, can provide the routine scheduling of maintenance automatically. Computer software programs are available to generate a daily inspection work order, tasking assignments for field personnel, and engineering information for budgetary considerations. The maintenance operations and frequency database can track maintenance items over periods ranging from daily to yearly. It is recommended that responsibility for this part of the program be specifically assigned. Among some of the advances in computer applications available are

- Daily inspection/maintenance work orders;
- System inventory;
- Historical record of system problems;
- Current system conditions for predictive management;
- State-of-the-system analysis for budgetary considerations on repairs;
- Deficiency inventory;
- Extent of rehabilitation/replacement work;
- Percent of system television inspection; and
- Television inspection findings, including
    - Active service connections,
    - Structural conditions,
    - Maintenance problems, and
    - Evaluation of extraneous flows.

In general, the staffing level and cost to support a computerized work order center can be justified because it improves scheduling and reviews of priorities for backlogged work. Computer systems can be used more easily to predict future needs for capital outlay, O&M funding levels, contracted or inhouse repair funding, and associated changes in required workforce or contractor skills. From such knowledge, training topics and education needs can be evaluated. Computer work orders can also be used for long-term scheduling to facilitate inflow and infiltration (I/I) metering, video inspection of the system, and industrial waste management.

Systematic approaches to system inventory and conditions typically require long-term benchmark procedures to provide a schedule of coverage. This process typically is separate from routine maintenance scheduling that requires short intervals between line cleaning inspections, although the same personnel may be used for both. Recorded inspection intervals of the system vary from agency to agency and are determined by the current condition of the system, system size, labor requirements, and equipment available. An example of long-term inspection benchmarking is given in Table 3.2.

While Table 3.2 may not fit specific circumstances, the goal of benchmarking is to provide an expanding inventory and condition database on the system, with regular observation of known defects. This type of program helps reduce the incidence of unanticipated failure, allowing minor repairs to be made before major repairs are necessary. From the system inventory and maintenance records, a site-specific benchmark schedule can be developed using computer-assisted hydraulic modeling and can be expanded as needed (see Chapter 8 for more information on benchmarking).

Table 3.2 Example of long-term inspection benchmarking.

Item	Checks	Frequency
Manholes	3–5 years (subject to flow conditions and lateral shift conditions)	
Sewer lines	Sound pipe, good flow characteristics	5–8 years
	Pipe with minor defects, good flow characteristics	3–5 years
	Pipe with minor defects, poor flow characteristics	1–3 years
	Pipe with significant defects, poor flow characteristics	Annually

INSPECTION. As stated previously, system inspection should be part of the daily activity of the maintenance staff. Everyone involved in field activity should gather information on system conditions as they perform their assigned tasks. Collection systems management is charged with the primary task of maintaining the system and the investment it represents. To do this properly, an inventory of the system, with regular updates, is necessary for planning and future community development. This inventory should include primary information on the system, including

- Population served by the system;
- Total system size in length of lines;
- Inventory of pipe lengths, sizes, and access points;
- System appurtenances, bypasses, siphons, overflows, diversions, and splits;
- System component installation dates and materials;
- Total theoretical capacity of the system;
- Total theoretical generation of wastewater by population;
- Volume customers;
- Industrial and commercial users with high discharge levels;
- Total current flow volume;
- Total length of off-road easements and rights of way;
- Force main locations and lengths;
- Summary of pipe sizes and lengths of each size;
- Number, location, and capacity of pumping stations;
- Number and location of siphons and inverted siphons;
- Number, location, and size of tide or flood gates;
- Equipment on hand (maintenance, excavating, and repair equipment); and
- Number of field personnel available.

After the primary inventory has been established, secondary information is added that describes the current conditions of the system. Because these conditions are constantly changing, this becomes the active list by which most maintenance and repair priorities are made. Secondary information should include

- Breakdown of pipe sizes and materials;
- Breakdown of interceptor, trunk, branch, and lateral lines maintained;
- Condition of physical line sections as available;
- Manhole inventory—location, size, type, and condition;
- Lift station components, including pumps and valves; and
- Complaint totals, including type, location, cause, and frequency.

These inventory lists contain the basic information needed to build a working systems inventory that can be expanded. There are many more detailed items that can be added, depending on the size, complexity, and topography of the system.

After basic primary and secondary inventory formats have been developed, daily inspection of the system will provide usable information and accelerate timely maintenance before serious problems develop through neglect.

As discussed later in this chapter, systems inspection should always identify extraneous water entry. In both surface and underground inspection, evidence of water entry should be noted for future consideration or solution. Adding this information to the system inventory will create a far more comprehensive database from which hydraulic modeling can be used to predict system upgrades or repair needs.

Surface Inspection. Some problems can be detected by visual inspection while walking or traveling along the rights-of-way of the sewer line, both in the street and in off-road easements. Sidewalk irregularities; cracked, settled, or dipped pavement; ponding over manholes; or depressions along the path of the pipe can be indicators of leaking or collapsing pipes. If a joint is bad or a pipe is broken, wastewater may wash away the surrounding soil and create a cavity beneath the surface. Sometimes, the weight of the overlying soil is enough to cause collapse and a depression at the surface. In easements, these depressions can be seen as flooded or sunken areas along the pipe route, especially after a significant wet-weather event.

Cavities under the road surface are often reported as a depression in the street. Such a report may go to a street repair crew, and the depression would be filled without further investigation. To avoid this, the cooperation of the street repair department is essential. Cavities that are left unattended will enlarge, and the weight of a vehicle can be enough to cause a road collapse. Depressions (settling) can also be the result of inadequate compaction during construction or subsequent utility work.

Manhole inspections are another source of surface surveillance information; problems can be reported on a standard form. The report form notes items such as evidence of high water or surcharging; missing bricks, concrete, or rungs; and water entry through the barrel, riser, frame, or cover of the manhole. Broken, raised, or sunken manhole frames or covers can be noted and repaired quickly.

Manholes periodically are examined for any sign of bedding material in the invert that suggests a breech or collapse in the upstream line. Deteriorated steps may indicate the presence of corrosive gases, possibly hydrogen sulfide. Information on the completed forms is transferred to the computer and loaded to the data file to be used later for listing priority, difficulty, or common traits.

Television Inspection. Some older methods of visually inspecting pipe, such as manhole entry to lamp portions of sewer pipes, entry into large pipe for visual inspections, and the use of mirrors on adjustable poles, are still used. However, most internal inspection is performed by the use of closed-circuit television camera systems. Televising of sewer pipes eliminates the necessity of personnel entry into manholes and can provide a videotaped record of the conditions in existing pipes.

Different types of cameras are available for different types of collection systems environments. There are two primary types of camera systems used for this work: cable-deployed cameras, which are pulled through a pipe section with remote winches from a truck or van; and self-propelled camera systems, which use an internally geared crawler or tractor that operates on a tread or wheels to move the camera forward or backward. These systems, typically referred to as "main-line" systems, can be extremely versatile in inspecting pipes 0.15 to 2 m (6 to 80 in.) in diameter. There are also explosion-proof versions of main-line systems for use in hazardous atmospheres. In addition to main-line systems, portable camera systems are used to televise pipe ranging in diameter from 0.1 to 0.6 m (4 to 24 in.).

Main-line camera equipment can be a simple, fixed-view camera or can incorporate a rotating head, typically referred to as *pan and tilt,* allowing a full radial view of joints and the ability to look into service connections. Main-line camera systems have capabilities to inspect pipes with diameters larger than 2 m (80 in.) by using special flotation devices and high-intensity lighting.

Main-line cameras can be used for 0.15 to 2 m (6 to 80 in.) diameter pipe and may have the capability to inspect larger pipes using flotation devices and high-intensity lighting. On the other end of the spectrum, inexpensive minicamera systems can be used for pipes from 0.15 to 0.05 m (6 to 2 in.) or less in diameter. These cameras are used with special push cable, pulled by cable, or propelled with flushing equipment to move the camera through the pipe. These small systems typically are used in small-pipe inspection (such as service lines) or to provide a spot diagnosis of a problem during an emergency.

The cost of this equipment has decreased enough for many agencies to use a camera regularly in system maintenance. Many contractors offer television inspection service for smaller agencies. The pull-type camera, color or black-and-white, is the most common type, with various sizes of skids for different pipe sizes. Other equipment includes lights, a pulley assembly with two winch stands, a single or multiconductor power and video cable, a cable reel with a location meter or footage encoder, a television monitor with a recorder, power and camera controls, and a service van or truck with a portable generator. Some systems are set up for chemical grouting from the same truck so that joints and leaking areas can be sealed. A high-velocity cleaner or a small hand or powered winch typically is used for pulling the cable between manholes to connect to the camera. The camera and cable are then connected, placed in the upstream manhole, and pulled downstream with the flow. This step can be eliminated when a self-propelled camera is used.

After the televising operation begins, the operator is responsible for noting conditions in the pipe. This is done using a video recording of the image with an audio commentary. In some instances, the operator may have a screen-displayed video log that can be also used for pipe condition notes or an onboard computer that allows simple keyed entries to a programmed data management system. The following list gives suggested procedures for accurate recording of televised pipe conditions:

- Enter data view information, such as the date and job number, after the camera is placed in the manhole.
- Announce or note why the pipe is being inspected and if tributary pipes are plugged or active, or flow is being bypassed.
- Record pipe material, pipe joint intervals, and pipe diameter.
- Record all structural defects as they are encountered.
- Record all extraneous flow sources and assign leakage rate.
- Identify all service connections and note whether they are active, capped, or abandoned.
- Identify maintenance problems such as roots, grease, sediments, or dips.
- Note whether the entire line was inspected, or note the reason for incomplete inspection.

Television inspection systems operators should be trained in equipment operation, sewer defect classifications, and data recording procedures. Any defects in the pipe should be noted on the recording tape, and the appropriate defect code should be noted on a standard form (see Figure 3.2). This form can be used either with a computer application or as a paper record form. These records should be documented according to the coding or rating system adopted by the investigating agency. There are numerous coding systems currently in use, ranging from basic handwritten codes with 10 to 20 code classifications to computer coding applications that contain numerous code classifications and matrixes. Appendix A lists examples of typical defects classification codes for use in visual inspection of collection systems. This notation should be used on all forms to standardize communication.

Computer applications permit the storage of inspection data in computer files for rapid processing and cross referencing capabilities to evaluate and rank large quantities of data. The agency should select a coding system that is appropriate for its particular applications and train operators to properly apply the coding system to avoid confusion or the potential of having to perform work a second time to accurately code defects. With a coding system, some agencies use a rating system used by the operator to help determine priority levels. If a rating scale is used, operators should be trained in making these assessments because most ratings will be subjective to some degree.

Data management is required after surveillance records are collected. Some agencies continue to use a paper system that splits information into areas of responsibility so that it can be reported to various departments. The information typically is logged into each department's records to be acted on in order of priority. With the increased use and capacity of computer systems, information can be downloaded to a database, sorted, and merged with each department's records more easily. This procedure takes less time and the data is handled only once. Priorities and needed repair or upgrading projects are clear so that long-term budgets can be used and crisis situations can be avoided.

Inflow and Infiltration Identification. An effective preventive maintenance program includes identification of extraneous flows in the collection system through inflow and infiltration (I/I). These extraneous flows can affect system performance and significantly increase the cost of wastewater treatment. Inflow and infiltration exist in all collection systems, and inspection crews are frequently tasked to identify and track I/I in parallel with inspection of the system condition.

Inflow is extraneous stormwater or snow melt runoff that enters a collection system through roof leaders; cleanouts; foundation drains; sump pumps; and cellar, yard, and area drains. Stormwater may also enter through older connections between a sanitary sewer and storm sewers and through defective manhole covers and frame seals.

Infiltration is water that enters a collection system from the ground through defective pipes, pipe joints, damaged service connections, or manhole walls. Infiltration most often is related to high groundwater levels but can also be influenced by storm events or leaking water mains. Storm-influenced infiltration typically occurs in shallow sewers, such as building service connections in porous soils.

Extraneous water introduction is often caused by wet-weather events or snow melting and can be seen during or just after wet weather. Inflow issues typically are easier to identify during wet weather because drains or sump pumps become active. These inflow sources are typically privately owned and need to be terminated by the property owner causing the discharge. In many instances, the property owners are not aware that they are discharging water to the collection system.

Infiltration can be more difficult to pinpoint and more expensive to resolve than inflow. Although the amount of infiltration can be estimated with some accuracy by a good flow-monitoring program, the actual infiltration sources can be especially difficult to find. They consist of underground defects that admit stormwater or snow melt that has soaked into the ground. Though some leaks can be found during and immediately after wet-weather events, it may sometimes take many days or weeks for the groundwater level to increase enough to contact the collection system. Flow monitoring typically is used to find these areas of extraneous flows, which are then televised during periods of high groundwater levels to determine where the flows originate. After being cataloged, repairs can be made on a priority schedule based on volume of groundwater entry. In some systems, I/I flows can account for the majority of the total wastewater flow that is treated, adding significantly to the cost of wastewater treatment.

The inspection crew tracks I/I flows in the system using testing techniques such as flow metering or portable instantaneous flow weirs to help identify or quantify extraneous flows. The use of closed-circuit television, smoke testing, and dyes can also provide information on extraneous flows. Some large agencies have found it economical to keep chemical grouting and sealing equipment in the unit with the closed-circuit television equipment for the inspection crew's use. Because it works closely with the general maintenance crew, the inspection crew typically is equipped with hand tools that allow the clearance of minor blockages and performance of daily checks.

Inflow and infiltration data collected by the inspection are used to identify and quantify specific sources of extraneous flows and determine segments of the system that do and do not have extraneous flows. After significant wet-weather flow metering and instantaneous weir measurement data have been recorded, the flow data are analyzed to determine where subsequent wet-weather and dry-weather I/I investigation work should be performed. The objective of the I/I investigation is the determination of extraneous flow sources so that appropriate action can be taken to reduce them.

I/I field investigations include the following tasks:

- Flow metering to measure sewer basin wastewater and extraneous flow rates and volumes;
- Manhole instantaneous weir readings, including
  - Determining wastewater and extraneous flow rates,
  - Determining sewer basins with excessive and nonexcessive extraneous flows,
  - Identifying sewer reaches with excessive extraneous flows, and

    — Quantifying manhole extraneous flows;

- Internal manhole inspection to identify extraneous flow sources in manholes and tributary pipes;
- Wet-weather television inspections to determine extraneous flow sources and distribution of extraneous flows between public sewers and private property sewers;
- Wet-weather surface inspections for manhole inflow sources;
- Dry-weather, closed-circuit television inspections;
- Dry-weather smoke testing;
- Dry- or wet-weather dye water testing at suspected storm sewer direct and indirect connections to the collection system;
- Building inspections;
- Identification of illegal or cross connections; and
- Small television camera inspections of service connections with extraneous flow or maintenance problems.

These investigations are necessary to control extraneous flows that hydraulically overload some sewers, pumping stations, and WWTPs. In most instances, severe extraneous flows will activate sanitary sewer overflows and cause backups that create health and environmental problems. The reduction of extraneous flows is beneficial because it lowers the cost of wastewater conveyance and treatment, increases system capacity available, and reduces the need for parallel sewers and flow equalization facilities.

Smoke Testing. The inspection crew, when evaluating I/I in the system, frequently uses smoke testing as a means to identify direct inflow connections. Smoke testing involves a specially mounted blower that forces air into a manhole and pipe, and smoke "bombs" of assorted duration. The smoke generated is nontoxic, has no odor, and typically is foggy white in color. The smoke is forced by the blower into the line and seeks the path of least resistance to exit the pipe. Under average circumstances, the structures in the section of pipe being tested will release this smoke from the vent stacks. If gutter drains are connected to the service line, smoke will appear from the downspouts. As a precaution, when engaging in smoke testing, descriptive information should be supplied to every building on the line before the testing. Most agencies use a simple door hanger that informs the user what is being done, allaying any concerns of toxicity of the smoke.

When conducting smoke testing operations, the fire department should be notified if there is a sump pump or floor drain connected to the service line through which the smoke will exit the well, possibly leading people to mistake the smoke for a fire. It is important to note that smoke testing typically is not effective in locating all sump pump connections. Some pump systems incorporate "P" traps that block the smoke, or are in basements that are not viewable from the street or lawn areas.

Some property owners are not aware of illegal connections to the system. In such cases, when smoke is viewed exiting from anything but the vent stack, a photo is taken and shown to the owner. Plans can then be made to terminate the connection, typically at the owner's expense. In some areas, local, state, or county funding may be available to help defray the cost to system users of disconnecting these sources.

Smoke testing can also show cross connections to storm sewers and other underground leak points. Smoke will exit from storm drains or appear out of the ground, indicating a problem. These problems will require visual inspection of the pipe to determine where the opening or cross connection is and can sometimes be difficult to trace because the smoke can drift between underground pockets, appearing some distance away from the actual release point.

Another means of inflow control is building inspections. Though time and manpower consuming, it may be the only way to find illegally connected sump pumps and floor drains. Building inspection can be conducted as smoke testing is ongoing, with few property owners not allowing access to structures. Some agencies work with the local fire and building departments when they inspect private and commercial property, asking them to note any pumps or obvious drainage that goes to the service connection.

Temporary flow-monitoring and data analysis of targeted areas of a collection system helps reveal the location of I/I.

FLOW MEASUREMENT. A fundamental of collection systems management is understanding how gravity flow and pumping stations convey wastewater flows through the system. Knowing how much water is moving through the system, how fast is it moving, and what portion of the volume should be based on the number and types of users can help determine how much of that volume is I/I. Several methods can be used to measure wastewater flows. The accuracy of these methods can be approximately $\pm$ 5% with electronic measuring equipment. The degree of required accuracy is determined by the purpose of flow measurement.

Flow measurements will determine many things, such as the location of areas having high rates of I/I during wet weather. In addition to revealing where extraneous flows are, flow measurements inform the system manager of the amount of wastewater being conveyed versus the capacity and estimated user volume of the system overall or in a given area. When development begins to tax the current capacities of pipes, flow measurement may be the only means available to regulate land use or system upgrade parameters. This information is also valuable for bypass pumping or sizing a new pump for a lift station. Flow measurements can be used to determine I/I by comparing dry- and wet-weather measurements. In some cases, flow measurements are used to detect wastewater leaking out of the pipe and water entering it.

The oldest method of flow measurement, the time required for an object to float between two points, can be used to approximate a given flow rate. The equation, flow = area $\times$ velocity, can be used to approximate flow rates. Alternatively, advances in flow metering and computer applications have produced a wide array of metering equipment that can be used to provide more accurate flow data. Some flow measurement methods include

- Continuous metering with and without primary measurement devices (flumes and weirs);
- Instantaneous flow measurements with calibrated weirs; and
- Velocity and depth methods, including
  - Velocity by meter,
  - Velocity by dye test, and
  - Velocity by floating object.

METERING. Continuous metering is used for a variety of different reasons such as measuring treatment flow rates, measuring flow rates from joining communities for billing purposes, and projecting extraneous flow rates and volumes. Permanent metering facilities are often used for this purpose, using metering equipment that can record flow rates. Older metering facilities initially used depth-only meters to measure flow rates, which can be inaccurate if submerged flow or backwater conditions occur or debris buildup affects the primary device. Typically, these meters were used in permanent or temporary manhole locations, providing a printed record on a circular chart on a daily or weekly basis. Advances in technology now provide meters that can read depth and velocity simultaneously, providing higher accuracy during normal and surcharged flow conditions over longer periods of time. Data typically are stored electronically, or a telemetry system can be used to transfer information to a remote monitoring site.

Temporary metering is used to determine various flow conditions, including

- Dry-weather wastewater flow rates;
- Influence of extraneous flows;
- Flow rates for computer modeling;
- Peak and average daily flows for lift station design, bypass pumping, system upgrade, and interceptor sewer design;
- Available capacity for development;
- Location of bottlenecks;
- Project exfiltration rates; and
- Sewers with maintenance or capacity difficulties.

Many agencies use flow-monitoring devices to measure wastewater flows from subbasins in their overall collection systems to assess the effects of I/I. Dry-weather, wet-weather, and rainfall-induced flow conditions are recorded. The dry-weather diurnal flow rates are used to confirm wastewater discharges and the presence of groundwater infiltration. Wet-weather flow conditions occur when groundwater is elevated but no significant rainfalls occur. During these conditions groundwater infiltration rates increase. The

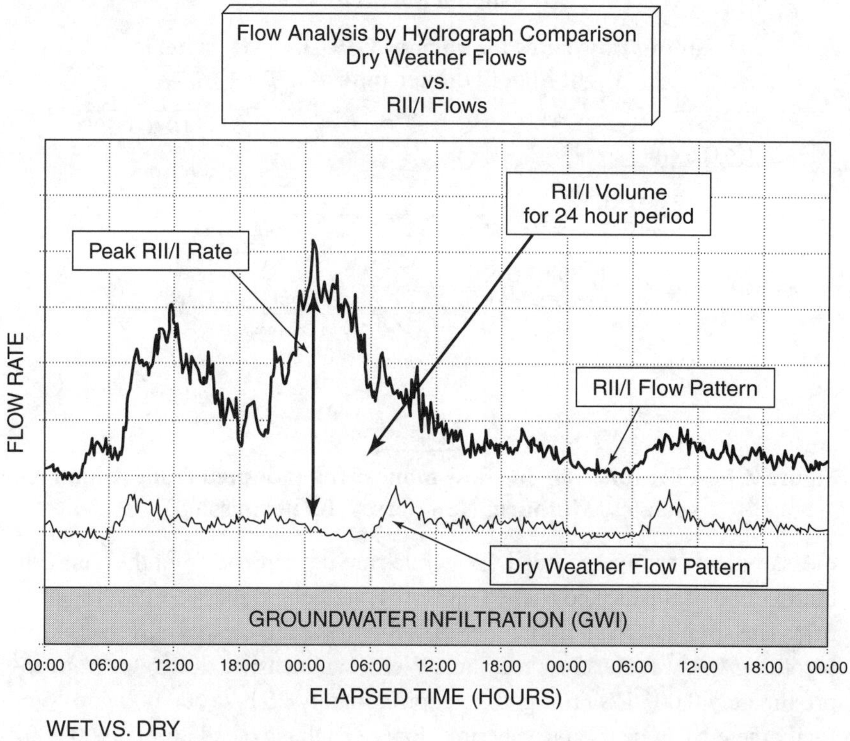

Figure 3.6 Flow analysis by hydrograph comparison; dry weather flows versus RII/I flows.

highest I/I rates occur during severe prolonged rainfalls, when rainfall inducted infiltration/inflow (RII/I) is activated. Flow-monitoring data can be used to determine RII/I flow rates and volumes.

Projected groundwater infiltration and RII/I rates are determined through numerical and graphical comparisons of dry- and wet-weather flows. A graphical comparison of dry- and wet-weather flow patterns is an excellent method of projecting I/I rates and visualizing the I/I impact at the monitoring site. Figure 3.6 is a hydrograph (a plot of wastewater flow during a selected period of time) that projects RII/I during a 24-hour period. A rainfall-induced flow plot is superimposed over a typical dry-weather flow pattern for the same day of the week (for example, Monday versus Monday) at the same metering site. The hydrograph notes the peak rate and volume of RII/I during the analysis period. Through this type of flow comparison, the daily flow patterns or cycles at a particular metering site are matched. Any difference in the flow patterns is attributed to wastewater fluctuations or I/I.

Instantaneous flow measurements are used for purposes similar to those of permanent or temporary flow-monitoring systems, but in smaller drainage basins or where continuous flow measurement is impractical. By inserting portable, calibrated weirs into manhole pipes, accurate instantaneous flow

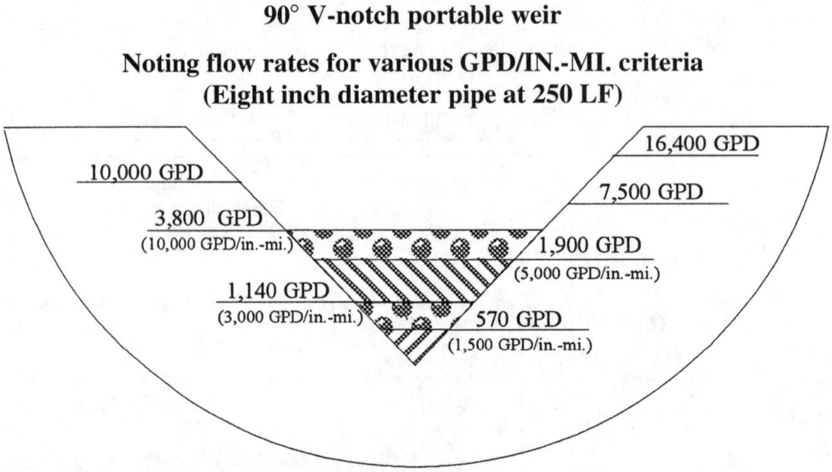

Figure 3.7 Portable weir for flow monitoring (adapted from Killam Associates, Millburn, New Jersey, by permission).

measurements can be recorded. Flow rates are determined from the weir calibration line nearest to the backed up flow level behind the weir (see Figure 3.7). The simplicity and portability of weirs make them especially useful for recording wet-weather flow readings. Weirs can measure flow rates from approximately 0.005 L/s (100 gpd) to approximately 26 L/s (600 000 gpd). Velocity meters can be used to measure flow rates that exceed 26 L/s (600 000 gpd), flows in very large pipes, and collection system flows in surcharged manholes. A depth of flow measurement is also needed for velocity measurements and typically is recorded with weir readings.

Along with flow monitoring of extraneous flows, rain gauges should also be used to determine rainfall amounts, duration, and intensity. Comparisons between rainfall and flow monitoring determine the relationship of rainfalls to collection system extraneous flow responses.

CLEANING

Maintaining clean sewers is an initial objective of a preventive maintenance program. Partial obstructions exist in collection systems because of intruding roots, grease and detergent accumulations, poorly constructed sewers, structural damage, vandalism, and poor design considerations. At some locations, sediment depositions and debris accumulations are created by design parameters, allowing sediments to fall out of suspension. These deposits of solids typically occur in sewers with mild slopes, where wastewater velocities are less than 0.6 m/s (2 ft/sec). Hydraulic bottlenecks where wastewater stalls can also be areas where solids deposition occurs. This is a typical problem in interceptor sewers, especially in newly constructed lines that are carrying only a fraction of their design flow. Extremely low velocities in these sewers sometimes permit settlement of organic solids that leads to odor problems. Odors can also be generated in steeply sloped sewers and at drop connections that cause turbulent flow conditions that release entrained gases from wastewater.

Hydraulic problems caused by design are difficult to correct after the system has been installed. The cost of replacement may be prohibitive or cause service interruptions if no bypass system is available (see Chapter 2). The design plans should be reviewed carefully for areas in which bottlenecks or low flow potential exists, and those problems should be addressed. If hydraulic problem areas exist, routine inspections and cleaning will be necessary to increase conveyance capacity. Combining system maintenance records with flow data may allow a forecast, or prediction, of when maintenance will be needed. Predictive management of these areas will help lessen the cost of manpower and the inconvenience to system users, and allow repairs to be planned and spread over time to lessen workload and costs.

Roots, grease, and deposited solids are the most typical cleaning problems. Effective control of these problems necessitates an understanding of how they develop. Grease builds up in a pipe over time as waste oils from food preparation float on the surface of the wastewater and coat the inside of the pipe. Repeated coatings can reduce the pipe capacity to a fraction of its original size and can inhibit flow. The grease coating hardens over time and becomes difficult to remove. This problem typically is found near restaurants and commercial food processors. Household garbage disposal units, however, also affect the character of residential wastewater and can lead to grease problems.

Roots can be destructive in collection systems and, if not controlled, can cause problems that necessitate major repair or reconstruction. Roots enter the system through cracks, bad joints, and service connections and can grow large enough to completely block flow. Root intrusion, in some instances, can allow groundwater infiltration during high-water-table periods by causing an increase in the gap in the pipe or pipe joint in areas where backfill is voided or easily moved.

Sediment problems typically are a result of a shallow or flat grade or of bedding materials entering the pipe from a pipe defect or service connection. Slow flows allow solids to fall out of suspension and gather on the bottom of the pipe. In larger lines (0.7 m [30 in.] and larger), compaction from the weight of water and material can make sediments difficult to remove. Mechanical failures such as a misaligned joint or collapse can result in heavier materials such as rocks, gravel, or sand in the line that can quickly build.

HYDRAULIC CLEANING. Hydraulic cleaning refers to any application of water to clean the pipe. Some hydraulic cleaning methods require the use of tools (sewer balls, scooters, and kites) and depend on the flow in the pipe to help the tool clean. High-velocity flushing machines typically are used to provide a combination of water pressure and volume to move materials to a removal point.

Sewer balls, scooters, or kites are often used for routine maintenance in pipes that have a long distance between access points, or where conventional

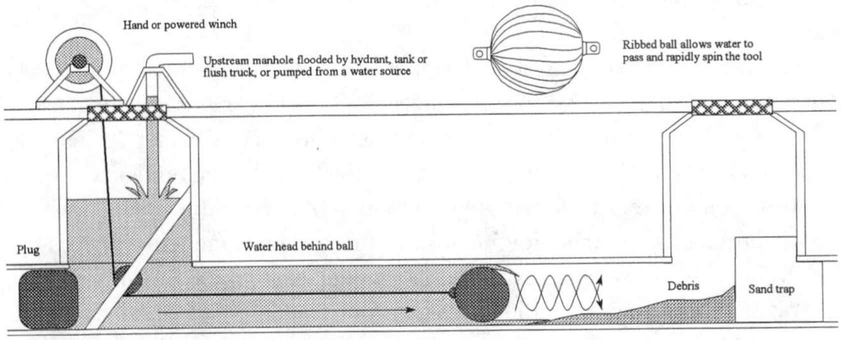

Figure 3.8 Sewer balling operation.

equipment cannot gain access, such as in bogs or thick brush easement areas. These cleaning methods require a trap at the downstream end to catch debris, a winch to control the tool's descent or travel down the pipe, and a water source to provide hydraulic head for the tool to operate. The tool is lowered into the upstream manhole and slowly winched downstream. The tool is spun or agitated by the head of water, which builds up behind it, dislodging and moving materials downstream to the trap.

Knowledge of the system and flow rates is important because water will back up behind the tool as it is moved down the pipe, and care must be taken to ensure that no service customer is flooded. Caution is needed when using sewer balls because protruding connections, severe misalignments, or heavy root intrusion or waste oil accumulations may prevent the ball from moving past the obstruction. This type of cleaning equipment is inexpensive and fairly easy to operate, although sewer balls larger than 0.6 m (24 in.) in diameter are less manageable. The use of sewer balls is illustrated in Figure 3.8.

Use of a hydrant, tank truck, or portable pumps to operate sewer balls or scooters may be necessary if the line flow is not sufficient to provide the motive force behind the tool. Some older systems were designed to use tank trucks or hydrants to flush the lines, incorporating flush boxes installed in key areas that allowed water to build up before the gate was lifted to allow water to pass into the next section. In this fashion, the same volume of water could be used for long distances in the system.

High-Velocity Flushing Machines. Hydraulic cleaning with a high-velocity cleaner is the most common technique used in basic system maintenance (Figure 3.9). This device is either trailer- or truck-mounted and has various sizes of water tanks and different types of power drives, depending on application.

Flushing machines are best suited to remove sediments and deposits such as light grease, sand, and collected solids in pipes up to approximately 75 cm (30-in.) in diameter. Flushers can have difficulty removing roots and hardened grease efficiently; however, they are often used in finding these areas

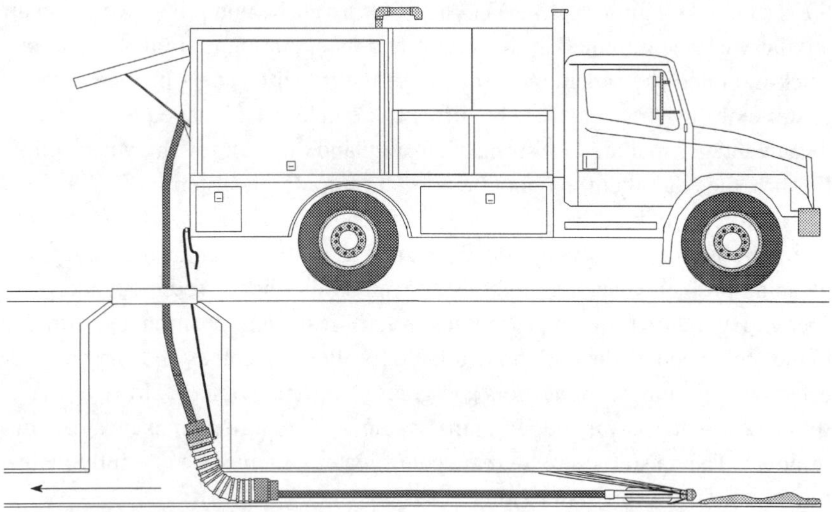

Figure 3.9 High-velocity cleaner.

for follow-up use of mechanical cleaning equipment. Additionally, as pipes increase in diameter, flushing machines will lose their effectiveness as the hose and nozzle meander in the line or the material depth becomes too great to be moved efficiently.

A water pump delivers water through a hose nozzle at a high pressure and volume, moving most materials in the pipe. The flow rates of the machines vary from 0.63 to 6.3 L/s (10 to 100 gpm), and pressures vary from approximately 10 340 to 27 600 kPa (1 500 to 4 000 psi). Typically, a machine operating at 4.1 L/s (65 gpm) and 13 800 kPa (2 000 psi) on 2.5-cm (1-in.) hose systems and 2.2 L/s (35 gpm) and 13 800 kPa (2 000 psi) on 2-cm (0.75-in.) hose systems is used. Combinations of pressure and volume can be varied for special problems.

Trailer- and truck-mounted styles have different uses. The truck-mounted, high-velocity cleaner has a water storage capacity of between approximately 3 800 and 13 300 L (1 000 and 3 500 gal). Using a portable reservoir saves refilling time. The pump can be powered by an auxiliary engine or other power takeoff option. Depending on the type of pump and power source, a pump can deliver from 4.1 to 6.3 L/s (65 to 100 gpm) at 13 800 to 20 700 kPa (2 000 to 3 000 psi). Truck-mounted cleaners are effective for most types of cleaning problems.

All necessary tools can be kept on the truck because a self-contained vehicle can more easily transport workers and equipment and facilitates high visibility. The disadvantages are that a truck typically costs more than trailer equipment, can consume higher volumes of potable water, and cannot be used if the chassis is not in running condition.

Trailer-mounted cleaners have water storage capacities from approximately 1 100 to 3 800 L (300 to 1 000 gal) and can deliver between 2.2 and

6 L/s (35 and 60 gpm) at 13 800 kPa (2 000 psi). The pump is powered by an auxiliary engine mounted in the trailer, and its operation is similar to that of truck-mounted equipment. Advantages of the trailer cleaners include lower cost, flexibility (transportable by different vehicles), and maneuverability. Disadvantages include less storage capacity and difficult towing when full. Further, the equipment may not be efficient for a large system.

High-velocity pipe-cleaning equipment can be equipped with a variety of nozzles used for different applications such as routine maintenance, specific cleaning problems (such as siphons), or removal of heavy sediment and grease. Hydraulic root cutters that use water pressure to position and turn a blade on the end of the tool are available for these machines and can be used effectively to remove light roots and grease from many pipes. However, these tools can become lodged in the system and require pipe excavation for removal. They also require a great deal of water and must be carefully maintained after use to prevent rusting.

Because flushing machines often will pass underneath, over, or around obstructions, a proofing tool can be used after cleaning operations to ensure the pipe is cleaned. A proofing tool is designed to pass within 25 mm (1 in.) of the pipe walls and will quickly show an obstruction in the pipe. The proofing tool attaches to the flusher hose in place of a fin extension for the nozzle. After the nozzle is mounted, it is propelled up the line until it clears the upstream manhole or encounters an obstruction. If an obstruction is encountered, it is noted, and a mechanical cleaner or camera can be dispatched to the site for removal or identification.

Air-Conveyance Systems. Many agencies include an air-conveyance system or a vacuum system mounted with the flusher to remove collected deposits and debris from a manhole or sump. This type of machine typically is referred to as a *combination cleaner*. These cleaners use the same cleaning techniques as high-velocity cleaners and use a vacuum system, powered with a positive-displacement or air pump, or with a series of fans, to pull material from a manhole or sump into a debris hopper. Connectable tubing with a nozzle attached is used to reach the debris. Material is vacuumed into a collector from which excess water can be siphoned and returned to the system.

The combination system typically is chassis-mounted to allow the vacuum and the high-velocity cleaner to work simultaneously. This arrangement allows its use in a wide variety of applications ranging from line maintenance to WWTP maintenance. The combination unit, however, is more expensive than a high-velocity cleaner (and can be more expensive than a high-velocity cleaner with a trailer vacuum unit). The truck-mounted unit requires a large chassis, which reduces maneuverability and typically requires a large storage area or garage during freezing weather.

Standalone vacuum systems can be either truck- or trailer-mounted, depending on the application needs. The trailer-mounted vacuum unit typically operates with a small positive-displacement pump, similar to truck-mounted equipment, but uses a smaller suction hose of between 0.1 and 0.15 m (4 and 6 in.) diameter and has a smaller storage capacity. The advantages of this kind of unit include lower cost and easier storage. In addition, the trailer can be towed by any vehicle, and the high-velocity cleaner and trailer vacuum can be separated and used in different areas at the same time. Disadvantages include smaller debris storage capacity, more "dump" time, and less capability than the truck-mounted unit for handling large workloads.

Truck-mounted vacuum systems, either standalone or mounted with a high-velocity flushing system, offer a higher range of air-conveyance power and debris capacities depending on the application need. These larger machines typically use either a positive-displacement pump or fans to move tremendous volumes of air. With the use of 0.15 to 0.2 m (6 to 8 in.) diameter connectable vacuum tubes, the debris storage can range from 5 to 11 m^3 (7 to 15 cu yd). The advantages of the truck-mounted vacuum over the trailer-mounted system are the increased capacity and capability of workloads, decreased downtime caused by dumping debris, and elimination of large equipment to be towed. Disadvantages are reduced maneuvering ability and increased cost at purchase.

When performing cleaning work with hydraulic cleaners in large-diameter pipes near treatment facilities, the potential to place solids in suspension is high. Treatment plant personnel should be notified of this work when possible to prepare for materials that may wash into the WWTP. Typically, treatment facilities have sumps located at the top end of the system that allow this material to settle for later removal.

MECHANICAL CLEANING. The term *mechanical cleaning* denotes the use of machinery to scrape, cut, or pull material out of a pipe. Among the most typical methods of mechanical cleaning are hand rodding, power rodding, and the use of bucket machines. Many problems are solved efficiently and quickly with mechanical cleaners.

Hand rodding is one of the oldest forms of mechanical cleaning, using sectional steel rods of various diameters and lengths. A tool is placed on the end of the rod, and the rod is pushed up the line until it reaches the obstruction. The rod is then turned to rotate the tool and remove the obstruction.

Hand-rodding equipment is inexpensive and easy to use. Some small agencies depend on hand-rodding equipment alone; other agencies use this equipment in easements and other areas where larger machines cannot gain access. In many agencies, all collection systems vehicles are outfitted with lengths of hand rods used to relieve blockages that are in or near the pipe openings in manholes.

Power-Rodding Machines. Power-rodding machines typically are trailer- or truck-mounted machines that store the rod on a reel, using a mechanism to push or pull the rod into the pipe while rotating the reel, causing a tool to turn in the pipe (see Figure 3.10). Rodding machines are available with vari-

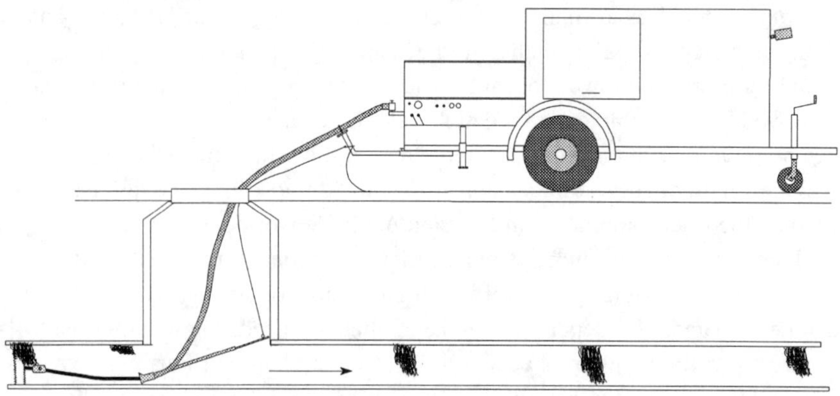

Figure 3.10 Sewer rodding machine.

ous engine sizes and offer a wide array of rod diameters and lengths in either sectional or continuous rod. Both types of rodders now offer automatic controls that can be set so that the machine functions with little operator effort.

Truck-mounted rodders offer power takeoff capability and the ability to carry equipment and personnel to the work site. Disadvantages of the truck-mounted unit include the increased cost of a truck purchase and limited maneuverability. Trailer-mounted units offer easier storage, lower initial cost, and greater maneuverability.

Rodders primarily are designed for use in clearing obstructions such as root intrusions, grease accumulations, rags and other stubborn materials, and stoppages. During the preventive maintenance process, the rodder can be used to clean an area of hardened grease quickly and inexpensively; the high-velocity machine can then be used on a regular basis to prevent buildup. In many agencies, the rodder is used for all emergency calls and to check lines that the flusher may have noted while using a proofing tool.

The rodder has advantages over most other pipe cleaners. Rodders can operate from a fixed point for as long as necessary, in virtually any weather condition. With the proper tools, tough roots in 0.1- to 0.4-m (4- to 15-in.) pipes can be removed. Except in the case of a collapsed line, rodders typically can eliminate stoppages regardless of the material. They are the least expensive equipment available for performing this type of work and have the lowest field operating costs, especially in areas where potable water sources are scarce or expensive to consume. Rodding machines do not work particularly well in debris removal, such as sediments or sand and gravel deposits. Setup time for rodders is slightly longer than for hydraulic cleaners, and greater operator skill is needed. Because the skill level required for this equipment is higher than for a hydraulic cleaner, safe and effective operation of the rodder requires thorough training. Experience is necessary to operate rods without damaging them.

Agencies with plastic pipe in their systems need special cleaning tools. A square-bar corkscrew is a rodding tool designed to remove obstructions without scouring the sides of the plastic pipe. After the debris has been bored and loosened, a flushing machine can be used to clean remnants. This method typically succeeds in plastic pipes, where root intrusion may not be a problem. Where plastic pipe is not visible from the manhole access as the result of repair, warning markers in the manholes can alert operators that regular mechanical cleaning tools may damage the pipe.

Rodding machines can proof the pipe by using a pipe-sized tool in the line. If the tool can pass through the pipe, the pipe is clean. Though most organic obstructions will yield to the rodding machine, protruding connections, badly misaligned joints, or other mechanical obstructions may not allow passage. If not, note the footage for future camera work that may be needed to identify the obstruction.

As with hydraulic cleaners, pipe proofing is essential to determine whether all materials have been removed. Records of proofing should be kept and entered to the master database for future reference.

Bucket Machines. Power bucket machines can be used in collection systems to remove roots, grease, and sediments from the pipe (Figure 3.11), but are specifically designed for large pipe debris removal. These machines are best used in pipes 0.75 m (30 in.) in diameter and larger, where large amounts of sediment and debris fall out of suspension. Over time, this material can compress and harden on the bottom of the pipe and become difficult to remove with conventional high-velocity cleaners and rodders. (As the diameter of the pipe increases, hydraulic cleaner nozzles and hoses will begin to meander in the pipe, moving materials to the sides instead of down the line.) These machines use heavy steel buckets designed to trap this debris,

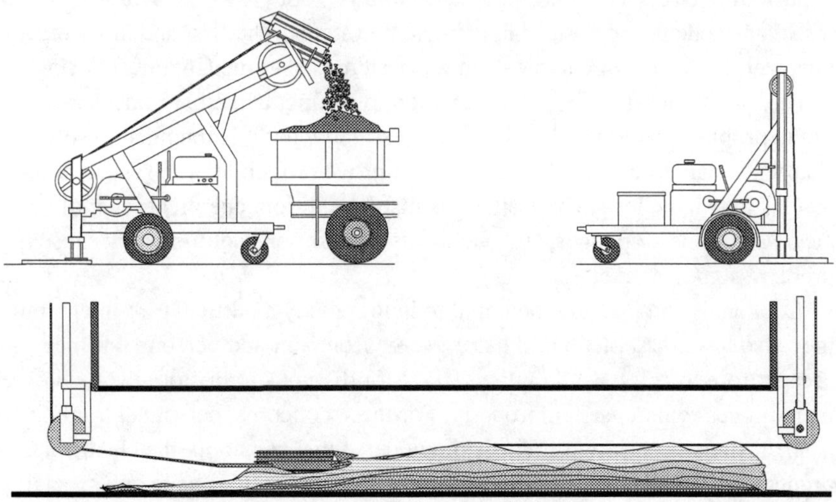

Figure 3.11 Sewer bucket machines.

bringing it to the surface for removal or loading into dump trucks. In extremely large pipes, special scraping tools can be used to gather the tremendous quantity of debris and move it to a point of retrieval.

Bucket machines are used in pairs: one machine pulls the bucket into the line and another pulls the bucket out of the line so that it can be emptied. This is an efficient way to remove material from a large-diameter pipe, especially if it is of substantial length (more than 250 m [800 ft]) between access points). Bucket machines typically are not practical for small-diameter (0.15- to 0.2-m [6- to 8-in.]) lines because protruding service connections and shifted or dropped pipes can cause the bucket to jam.

Bucket machines are available in a wide range of power options (12 to 75 kW [16 to 100 hp]) and are mechanically or hydraulically driven. They can clean lines ranging from 20 cm (8 in.) in diameter to the largest in the system and be adapted to specialized situations. Although the advent of vacuum equipment has limited their use, they are still the most efficient means of materials removal in large sewers. Bucket machines can also be used for slip lining or sleeving projects if the agency is engaged in this type of trenchless repair activity. As with rodding machines, these machines take a higher degree of skill to operate, and operator training is necessary.

Bucket machines are often used in large truck or interceptor lines near the WWTP. If this work is planned, notifying the WWTP will allow it to prepare, or look for materials that may become suspended and enter the plant. Typically, treatment facilities will have sumps to allow this material to fall out of suspension for later removal. However, if the sumps are full, or fill up with materials, help may be needed to evacuate the sump to prevent debris from entering the plant system.

CHEMICAL APPLICATIONS. Chemical applications can facilitate the control of odors, waste oils and grease buildup, root growth, corrosion, and insect and rodent populations in collection systems. The first and last word, however, in the use of chemicals in the system is caution. Chemical dosing can be an option after careful observation, planning, and close consideration of the problems associated with the process and what it is intended to solve. Elaborate claims for chemicals that are unproven for collection system applications may have less than desired results. In addition, chemicals cannot clear sewer line stoppages, and chemicals used for one solution may cause a problem somewhere else.

Chemicals can have the potential to help remedy a particular problem, but they also have the potential to harm the environment and personnel if they escape from the pipe, spill on the surface, or disrupt (if not stop) the treatment process when used improperly. For these concerns, personnel training is an absolute necessity if in fact agency personnel can apply the chemical product. Be sure to check with local or regional authorities as to any special licensing requirements that may be needed to apply a specific chemical. Some chemical processes may only be available through outside contractors with special licensing or equipment.

When using chemicals in the collection system, it is advisable to notify the treatment facility of the intended work. Some chemicals can quickly kill organisms used in wastewater treatment depending on the amount of the intended dosage or the distance between the dosing site and the treatment facility. In most instances, WWTP personnel can determine whether the dosing will pose a potential problem and offer suggestions on how to apply the chemical without disrupting critical plant systems.

Odor Control. Odorous compounds may be present in wastewater or generated by bacterial action, typically under anaerobic conditions, or from industrial discharges that can react with common wastewater ingredients. Hydrogen sulfide, a major odor source in wastewater systems, can be extremely dangerous. At low concentrations (5 to 40 mg/L [ppm]), hydrogen sulfide smells like rotten eggs; at high concentrations (100 mg/L [ppm] or more), the sense of smell is rapidly lost and may mislead an operator into thinking that conditions are safe. (See Chapter 1 for information on confined space procedures to be used to decrease this hazard and other hazards associated with entering confined spaces.) Sluggish flow, lengthy detention times between pump cycles at pumping stations, high temperatures, and low pH can all lead to high hydrogen sulfide levels. High concentrations can develop where ventilation is poor, such as in deep or low-sloped pipes or pumping stations.

Odor problems related to the presence of anaerobic bacteria can arise when wastewater is discharged from force mains to gravity sewers above the water line, when the speed of flow is in excess of 3 m/s (10 ft/s), or when drop manhole connections allow wastewater to fall into the receiving pipe, splashing the wastewater and releasing gases.

Odors from these causes typically are chronic, resulting in several calls from the same area or address. Often, a hydraulic cause can be found near the complaint site by looking in the upstream and downstream manholes. Odors can be released through structure vents and drift away from the immediate area, creating a false indication of a problem. Odor complaints can be tracked to locate potential hydraulic sources.

Most agencies will attempt to find a mechanical solution to these problems that offers a more permanent solution than chemical dosing. Redesign of a force main to discharge below the water level of the receiving pipe can often relieve turbulence-related gas release. Designed weirs can be installed in manholes along steep grades to decrease the momentum speed of the wastewater, reducing turbulence at the bottom of the grade.

In many cases, odor-causing conditions can be relieved with cleaning and regular maintenance. Low-flow, low-slope lines often allow organic material to be deposited and decompose. Routine cleaning of these areas can remove anaerobic odor sources and improve the flow characteristics to reduce their buildup.

If a permanent mechanical solution is impractical, or cannot be initiated quickly, chemical treatments can be used for temporary or long-term relief. Masking agents laced with various scents can help eliminate odor complaints. Masking agents can be applied to the flow itself, into the air via an atomizer, or as deodorants hung in manholes. Masking agents in many cases can hide the odor or blend with it to alter its offensiveness. Masking agents do not, however, eliminate the cause of the odor: they merely hide it. When hydrogen sulfide is a problem, the use of a masking agent is dangerous because gas can build to toxic levels undetected.

Aeration, dosing with hydrogen peroxide, and caustic addition by slug dose are effective in eliminating hydrogen sulfide gas. In sewer lines, chemical treatment of hydrogen sulfide is sometimes effective for short periods. Zinc, aluminum, and iron salts are used to precipitate the sulfides. Bactericides can be used to kill or inactivate anaerobic bacteria. Lime and caustic soda raise the pH of the wastewater and provide bactericidal action, depending on the conditions and quantity added, for periods ranging from 1 day to 2 weeks. Lime use, however, can create large amounts of lime sludge and a consequent disposal problem. Caustic soda or sodium hydroxide may also raise the pH of the wastewater and present problems for the treatment process, especially if added close to the WWTP or in heavy doses. As a general rule, the WWTP should be consulted when the addition of any chemicals to wastewater is being considered.

Chlorine can be injected to the system at various points. Pumping stations, regulators, or other locations at which water and power are available to operate injectors are likely locations for chlorine injection. Although chlorine is effective in hydrogen sulfide control, there are some disadvantages to it. The use of chlorine for odor control requires careful evaluation and is not considered a good remedy for long-term use. Special containment structures must be built for storing the chlorine. Doses of chlorine gas or hypochlorite solutions of 10 to 20 mg/L are toxic to most living organisms, including humans. Chlorine is a highly corrosive, hazardous material and may form toxic byproducts with the organic compounds in raw wastewater or with industrial discharges. It can pose as great a hazard as hydrogen sulfide and, in large doses, can kill aerobic bacteria and thus harm the treatment process. Chlorine accidents have the potential to significantly affect public safety.

Hydrogen peroxide, an oxidizer, has also proven helpful in odor control. Moderate doses of hydrogen peroxide have no effect on slime. A conditioning program is necessary to inactivate slime before dosing. The application program varies the dosage to account for fluctuating flow. Unlike chlorine, excessive hydrogen peroxide is not detrimental to the treatment process. However, the chemical can be expensive and concentrated solutions are potentially hazardous. Other chemical additives are effective but may be impractical. Potassium permanganate and ozone can be used successfully but may not be cost effective.

Maintaining clean pipe walls through proper maintenance procedures can be an effective odor control measure. High-velocity cleaning is sometimes used to control grease and slime buildup. Pulling a rubber squeegee with a bucket machine is also effective in removing the slimes or materials that generates gases. (For a more complete discussion of odor generation and control, see *Operation of Municipal Wastewater Treatment Plants* [WEF, 1990], *Odor Control for Wastewater Facilities* [WPCF, 1979], and *Sulfides in Wastewater Collection and Treatment Systems* [ASCE, 1989].)

Insect and Rodent Control. Collection systems provide a potentially ideal environment for insects and rodents. Insects feed off the materials found in wastewater and can eventually make their way through the system into structures to nest. Rodents use the collection system as a roadway, avoiding detection from predators and moving freely from one area to another. One cause of rodents in collection systems is the failure to effectively close off service connections in abandoned or razed buildings. Colonies of rodents can successfully avoid extermination through the collection system, building a new colony easily when discovered.

Insects and rodents typically are controlled in close cooperation with the public health agency or the vector control department. These agencies typically determine the best eradication method. Tight collection systems with proper fixture traps and connections, sealed joints, plugged unused building sewers, and good-quality manholes have few severe rodent problems. If rats do gain access to the system (for example, through open pipes or abandoned sewer connections, new construction, or repair areas), food supply from domestic garbage disposals alone can potentially support a large population.

Regular cleaning, flushing, and spraying of manholes or line section dosing with insecticide typically can control insect infestation. Vermin control can be more difficult because the colony may be difficult to locate. As with all chemicals, unless personnel are specifically trained in the application of the pest control chemicals used, it is often best to contract out this service or request assistance from the public health agency if possible.

Root Control. Chemical root control is an alternative to mechanical removal methods. Chemical treatment is widely understood to mean the application of products using sodium hydroxide, which may kill the root with repeated applications but will not necessarily inhibit regrowth. It is best to use only proven chemicals for which explicit instructions are available, and to follow these instructions carefully. Television monitoring of the sewer pipe is useful for determining the condition of the pipe, severity of the problem, best application method, and results after the process.

Chemical dosing of roots typically is more expensive than conventional removal methods using a rodding machine or hydraulic root-cutting tool.

However, it can be an effective solution in areas with high levels of traffic that do not allow safe machine operation, in low-impact areas such as easements through landscaped yards, when conventional removal methods are not practical, or when root diameter continues to expand and threaten the pipe or pipe joint.

Proper application of an effective herbicide ensures that roots in the pipe are killed, become brittle, and are eventually pulled away by the flow in the pipe. Several applications may be necessary for substantial root intrusions. The larger root masses are removed from the pipe to prevent blockages. With proper application, herbicides can inhibit root growth for 2 to 7 years.

Some of the newer products on the market use a combination ingredient such as sodium methyldithiocarbamate (metam sodium) 2,6-dichlorobenzonitrile (dichlobenil) and are applied in a foam that coats the pipe walls. Perhaps the best way to determine the success of the chemicals on a particular problem is to test them in specific areas. The chemicals can be used on various problems, and cost-effectiveness can be determined without a large financial outlay. Another way to determine the success of a chemical program is to talk with operators from other collection systems about their experiences.

Labels in manholes inform collection systems workers of the presence of chemicals. It is important to determine whether the treated pipes are in an area with a high water table. Many agents depend on translocation to kill the roots away from the pipe. Because these chemicals have not been used over long periods, no long-term data exist on the potential for introduction to drinking water sources from collection systems applications. Although manufacturers claim chemical agents are safe, some agencies are concerned about the potential liabilities associated with their use, and select application sites carefully.

It is important that WWTP operators be aware of all chemicals (and their ingredients) being used in the collection system so that problems are avoided at the WWTP. Chemical agents for root removal can be highly corrosive. Surface spills can occur, for example, from careless handling or traffic accidents; thus, spill containment planning and emergency procedures and equipment should be considered.

Grease Control. The use of additives for grease control has become popular during the last few years. Agents such as bacterial cultures, enzymes, hydroxides, caustics, bioacids, and neutralizers are available to help control severe grease buildups. Although these agents are sometimes effective in removing large amounts of solidified material in the pipe, problems can recur without regular application or preventive cleaning operations.

Many systems have incorporated grease trap requirements for commercial food establishments or processors that discharge a large volume of waste oils or tallow. Although legislation may require the installation of these traps, their routine maintenance and inspection can sometimes be lacking. Lack of maintenance on these traps eventually can negate their design, allowing waste oils into the collection system. These waste oils build up on the interior of the pipe in layers and harden over time. In some instances, mechanical cleaners are needed to cut these deposits from the pipe walls to reopen the full diameter. The local health department can be a valuable ally in getting owners to routinely maintain the traps and reduce the amount of waste oils they discharge to the system.

While grease traps, if maintained properly, can be an effective remedy for commercial enterprises, the advent of the household garbage disposal has created grease problems in residential areas that can cause damaging flow interruptions. Again, mechanical cleaners may be the only tool that can break up these hardened deposits, with routine flushing equipment used on a scheduled basis to prevent future buildup.

Chemical applications for grease removal can also be effective, especially in limited-access or high-traffic areas where machinery setup can be difficult or dangerous. If chemicals are a viable and affordable solution to a specific grease problem, proper care should be taken to ensure that no downstream workers are injured and that the chemical is not released to the environment. Also, rising gases from chemical reactions venting through the manhole can be dangerous, especially in confined spaces. Chemical applications should be strictly controlled by the field supervisor or local health department, with proper training supplied to all personnel who will be involved with the process.

BIOLOGICAL APPLICATIONS. Biological applications differ from chemical dosing, requiring the use of biological actions in place of chemical reaction to control grease, odor, or sulfide. Bioaugmentation is the addition of biological organisms to wastewater for constituent removal. Grease control via bioaugmentation and odor control via the addition of nitrate salts are growing biological applications as the performance of biological elements used in wastewater applications improves. The use of nitrate salts for hydrogen sulfide control is a different process from the use of strong oxidants such as peroxide or ferrous chloride, incorporating biological means to control the formation of sulfides in wastewater.

As with chemical applications, caution should be used when considering biological solutions to collection system problems. Always consult with WWTP personnel on the effects of introducing biological elements to wastewater because they may affect organisms in the WWTP. Human contact with biological products should be reviewed for potential hazards, and training should be mandatory for anyone handling these agents.

STOPPAGES AND OBSTRUCTIONS. In collection systems operation, stoppages are inevitable at some point, despite the best preventive maintenance program. The range of items, for example, that system users manage to get into pipes is surprising. From garments and soil to engines and tires, foreign materials will eventually find their way into the system. The key for a proactive effort to reduce the presence of obstructions, beyond those resulting from deliberate vandalism, is understanding where problems have been in the past and taking steps to relieve the buildup of materials that impede flow. Using records from past stoppages, routine maintenance should be scheduled to remove materials before they can accumulate and cause interruption in the flow.

Another source of obstructions is construction activity from nearby utilities or structures. Existing sewers previously inspected and found to be in excellent condition can be structurally damaged by excavation and boring activities such as road repairs or utility pole installation. In many instances, sewer pipes can be unknowingly damaged by construction or installation of foundation pilings, guardrails, or other underground utility conduits. Sometimes, unacceptable repairs are made, without notification to the utility, that eventually fail or allow pipe bedding material to enter the system.

Stoppages can result from a single cause or from multiple causes that alone would not present a problem. Construction in the system can introduce a wide variety of foreign materials such as concrete, wood scraps, bricks and gravel, excavation spoil, pails, and tools. Unmapped manhole structures and service connections that protrude into the pipe can be a source of flow interruption. Root intrusion, grease deposits, sediments falling out of suspension in low-flow lines, sharp edges or misaligned pipes, captured rags and paper masses, and wet-weather events flooding the system and carrying outside deposits are contributors to these blockages.

Although there is no sure remedy for predicting these stoppages, there are steps collection system managers can take to reduce them. An effective preventive maintenance program will reduce the largest percentage of possible blockages. Beyond this, other measures include

- Accurately mapping the system to determine hydraulic bottlenecks;
- Locking manhole covers in easement areas to prevent vandalism;
- Permitting local contractors before they begin excavation work on sewers;
- Requiring sanitary sewer markouts before construction work can begin near existing sewers;
- Requiring grease trap owners to maintain traps regularly; and
- Creating simple brochures for end users detailing what not to discharge to the collection system.

When a stoppage occurs, methods for clearing it depend on the nature, cause, and location of the problem. Line stoppages can be caused by any number of circumstances such as collapsed lines, physical objects in the pipe, root intrusion, grease or detergent buildup, faulty service connections, or misaligned joints. Table 3.3 lists various stoppage problems and some suggestions on how to relieve them. The key to removing stoppages is get-

Table 3.3 Types of pipe conditions and methods of removal.

Type of blockage material	Effective methods	Requisite condition	Remarks
Grease and waste oils	High-velocity flushing	In pipes up to 75 cm (30 in.) diameter	Effective if properly done and grease is soft
Grease and waste oils	Power rodding	In pipes up to 37.5 cm (15 in.) diameter	Effective on hardened grease and chronic problems
Grease and waste oils	Chemical/biological dosing	Waste must be of known composition	Can be effective in small areas; can be more expensive
Roots	Power rodding	In pipes up to 37.5 cm (15 in.) diameter	Effective; leaves loosened materials in line
Roots	High velocity flushing with hydraulic root cutter	In pipes up to 37.5 cm (15 in.) diameter	Can be effective on small roots; less effective on large roots
Roots	Chemical dosing	Application based on pipe size and root mass	Can kill roots over long term; too slow for emergency work; can be expensive
Sludge, sand, gravel, debris	High-velocity flushing/vacuuming	In pipes up to 75 cm (30 in.) diameter	Effective on static materials; can be lifted from manhole with vacuum; less effective in larger pipes
Sludge, sand, gravel, debris	Bucket machine	In pipes 37.5 cm (15 in.) diameter and over	Effective for removing static debris, especially in large pipe

(Continued)

Table 3.3 Types of pipe conditions and methods of removal (continued).

Type of blockage material	Effective methods	Requisite condition	Remarks
Sludge, sand, gravel, debris	Balling, scooters, kites	In pipes up to 60 cm (24 in.) diameter Proper pipe alignment[a]	Effective in moving static deposits, especially in long reach or unimproved access areas
Foreign objects, rocks, metal	Power rodding	In pipes up to 37.5 cm (15 in.) diameter	Effective in pushing/pulling objects to manhole access
Foreign objects, rocks, metal	High-velocity flushing	In pipes up to 37.5 cm (15 in.) diameter	Can be effective in moving with water pressure/flow; can be time consuming

[a] May require root removal before operation

ting the necessary equipment and trained personnel to the area as soon as possible.

Regardless of the type of cleaner used, mechanical or hydraulic, stoppages are best removed by accessing the downstream, dry manhole and sending the cleaning tool up the line to the source of the problem. When using hydraulic cleaners, operation from the surcharged manhole should be avoided at all costs. Getting the nozzle into the pipe is difficult; knowing it is safely in the pipe is not possible. If the nozzle is not in the pipe, it may point itself up the manhole when pressure is applied. When the nozzle is surrounded by water, it can move at an incredibly high rate of speed, exiting the surface of the surcharged manhole without warning, potentially injuring anyone in its path. Rodding machines may be used in a surcharged manhole with a higher degree of confidence; however, use of this practice should be limited to experienced operators.

If the problem occurs outside typical working hours, it may be advisable to relieve the stoppage and return the following day to clean the line fully with the proper tools. Many agencies prefer to use the rodding machine first if the stoppage is suspected to be roots, grease, or a physical obstruction. In most instances, this machine can relieve the obstruction and the source of the problem. The high-velocity cleaner can be used, but its effectiveness may be limited if the stoppage is caused by roots or large objects. If the stoppage is stubborn, the high-velocity cleaner may have to be refilled off site, adding time to an emergency job.

For each stoppage, the location, probable cause, equipment used, and time spent should be documented and used later in the collection systems profile. Thus, repeated problems might be prevented before an emergency occurs. Accurate recordkeeping for all preventive maintenance and emergency operations is a necessity to avoid successful claims litigation against the agency. All personnel that have the potential to respond to emergencies should be trained on how to quickly relieve the blockage and respond to affected users until a supervisor can reach the site. Legal claims against an agency can be successful because of a lack of good records and a casual statement by field personnel trying to be helpful to a distraught homeowner (see Chapter 5 for further information).

Emergency response is considered the highest profile activity in collection systems operation. It represents one of the few situations in which people, especially those affected by a blockage, notice the activity of a response team. People affected by stoppages typically are in a highly acute emotional state, especially if property has been damaged. The best procedure for response will result in quick alleviation of the problem, a professional appearance to the affected people, and a reduced likelihood of litigation.

While there are many different policies and procedures for responding to emergency calls, there are a few basic steps that should be taken to ensure the best possible outcome of any stoppage-related emergency. The basic procedures should be followed regardless of whether the call comes during the normal workday or after hours.

Responding personnel should immediately inspect the closest manhole to the affected locations. If dry, or running normally, move further upstream to determine if the line is blocked. If so, the equipment should be immediately set up to clean the line and remove the blockage. Appropriate traffic-control equipment such as emergency lights, road cones, and reflective vests should be used, especially at night. If no evidence of a main-line stoppage is found, the line should be cleaned as a routine procedure in the event there is an object causing an intermittent blockage in the line, and the affected property owners should be told of the situation.

Preferably, supervisory personnel should speak with the affected property owners. All emergency response crews should be trained not to comment on responsibility or fault, or make other statements until all the facts are known. The supervisor on site should have a camera to document any damage caused to property for future reference, and explain agency policy as it relates to cleanup and damage claims. Be polite and concerned when speaking to the affected property owners, keeping in mind their emotional state. Though not a normal occurrence, if threatened with legal action, immediately stop any further conversation with the involved parties and refer them to upper management or corporation counsel.

FORCE MAIN CLEANING. Unless the system is very small, or very unique, all collection systems incorporate lift stations that move wastewater either up a grade or through a long length of pipe to a receiving point. The pipes hold the pumped wastewater until it is discharged to a gravity sewer or WWTP. Force mains range in length from just a few feet (metres), for achieving a change of grade, to several miles (kilometres) long for transporting wastewater from a remote community to a treatment facility or system. Force mains are frequently constructed with iron piping to withstand the pressure of the water as it is pushed by pumps. Over time, the pipe may develop rust and scale on the interior surfaces that collect organic and solid materials, eventually causing the potential for stoppages, organic breakdown causing odors, or pipe failure from corrosion. Standard force main maintenance practices include pumping station equipment inspection (see Chapter 4) and maintenance of air-release valves to avoid reductions in pumping capacity.

To clean a force main, special tools are needed, and a special access called a "launch" tube needs to be installed. The launch tube typically looks like a "Y" strainer, working similar to an air lock. Keeping the bottom closed, the top can be opened to insert the cleaning tool, then resealed. After being secured, the bottom is opened and the tool is introduced to the force main.

The tool used to clean the force main is called a "pig." This device is shaped like a fat bullet, with an abrasive coating applied to the outside surface. The "pig" is made of a pliable polyester material to reduce the potential for getting caught in the line, and it is fairly lightweight, allowing the water to push it through the main to the far end. The name for this tool is derived from the sound it makes as it moves through the pipe, scraping the materials out.

The cleaning procedure typically begins with a small "pig" inserted in the launch tube. Released into the force main, it is retrieved when it appears at the far end, typically bringing with it an impressive load of dislodged deposits. After the small "pig" has cleared the line, successively larger "pigs" are used until the pipe has been scoured.

One of the complications of force main cleaning is the lack of a launch tube installed during initial construction. Many force mains require the construction of this launch tube to enable the use of the tool, which can add considerably to the cost. Some agencies lack the expertise or tools necessary to perform this cleaning operation and prefer to contract the operation to an experienced contractor who can handle most any situations that develop, from a stuck tool to a launch tube sealing failure. Because the force main is difficult to retrieve tools from, caution is imperative. It is recommended that agencies with no experience in performing this work talk with other regional agencies to see how they engage in this operation. In some instances, a sister agency may send experienced personnel to visit during the operation in exchange for a future favor. Engineering or contractors can also be an excellent source of information for this procedure.

Lifespans for force mains can be extremely difficult to determine. In many instances, a failure of the pipe is the only indicator that it is time to replace the line. The use of predictive management can provide a general time frame for potential replacement by including records of the date of installation, the type of pipe material used, the predicted lifespan of the pipe materials, and levels of known sulfides or corrosive discharges that can affect the pipe. Physical inspection of the force main may be possible with video cameras during off cycles if the pipe is not too long. The discharge end of the force main may have higher corrosion levels because it is not running full at this point. At this point, the force main interior may be visible, permitting an estimate of condition to be developed. From the video imaging references or physical observations, combined with the known physical properties of the pipe materials and wastewater characteristics, a lifespan estimate may be determined.

RECORDS AND REPORTS

As stated previously, there are three distinct methods of managing a collection system: crisis or reactive management, proactive or preventive management, and predictive management. Reactive management seldom incoporates records for future use; however, both proactive and predictive management require good records to develop a workable program.

Proper recording of collection systems data is necessary for a variety of reasons, including preventive management, budgetary considerations, litigation interests, and system inventory and analysis. Good records are required to effectively calculate costs, determine repair priorities, issue work orders, schedule repairs, and monitor daily operations.

Records range from the most elementary handwritten documents to computer-generated reports. Completeness and accuracy are critical when records are used as evidence in court, during hearings and budgeting, and in operations that require excavation. It is important that all employees involved in record preparation at any level be sure their contributions are as complete and accurate as possible. Computers can help keep this information current and accessible.

Computer-assisted applications will allow data to be merged in a wide variety of formats to accommodate each department. From automatic work order generation based on timeline information, to hydraulic modeling to determine system capacity and community or industrial growth, it is the timely recordkeeping discipline that will ultimately produce success or failure (see Chapter 8 for additional information on records).

COLLECTION SYSTEMS MAPS. One of the most typical problems in collection systems management is determining the locations of sewer lines and manholes. This is best done by keeping maps showing the locations of the sewer lines, cleanouts, manholes, and Y branches; the depth to the sewer invert; pipe material and size; and the direction of flow. Pumping stations, appurtenances, and WWTPs should be shown clearly and identified. System mapping for many agencies consists of large paper maps divided into overlapping, large-scale sections. These maps or sheets are bound into books that can be stored easily and taken into the field as needed. The use of microfilm records has made paper maps unnecessary in most cases. However, some agencies have both microfilm maps and large books available. For older systems, original maps may not be accurate because construction over the years can relocate entire sections of the system. If not maintained properly, maps can be quickly outdated by development and their accuracy reduced.

Developments in computer mapping make it possible to detail systems and add new structures easily. Computer-assisted drawing (CAD) system applications have expanded their parameters and offer a wide variety of uses, from simple mapping to computer-aided modeling for predictive management. Among some of these parameters are the ability to provide

- Information layering and dimensional imaging for use in rights-of-way;
- Variety of maps for different collection system applications or designs;

- Easily revised maps;
- Record of activity sites and causes of problems;
- Record of current and future rehabilitation sites;
- Record of current and projected maintenance sites; and
- Record of television inspection sites.

Programs, termed geographic information systems (GIS), range in scope and cost and are available through various specialty services. The GIS programs use global positioning satellites to establish coordinates for manhole locations, system alignment, and easement areas without permanent reference markers. This technology should be considered if the agency is using or planning to use computer software for information management. These integrated mapping programs can range from simple and relatively inexpensive to highly complex and expensive, depending on the size of the system and what management wishes to do with it. Many packages can be upgraded as the users' needs grow beyond the initial software, and integrate well with CAD and other operation software. User-defined attributes can be linked to mapping, monitoring, work order generation, and other aspects of collection systems operation.

MAINTENANCE RECORDS. An effective maintenance program includes a simple system for recording pertinent data. Maintaining records serve two purposes: to help record work done in the system, and to help plan future projects. Forms typically are prepared for all field crews to use. These forms include information such as the date, location, work performed, distance, reason for work, material removed, equipment used, condition of manholes, and any additional remarks.

In agencies that do not use computers, a card file system can be adopted to record collection systems maintenance data. A card could be used to record pertinent data describing the system and the maintenance performed. This card system is a concise record of work performed, and as time passes and the report grows, sources of recurring trouble, such as grease development or root intrusion, become evident. This information can then be transferred to a separate card file that lists the necessary preventive work. As necessary operations are done each month, the card is rotated to the rear of the monthly file. Ideally, some type of preventive maintenance activity is done each week, and the master record is amended to reflect trouble spots. The severity of the problem determines the frequency of cleaning or repair. Eventually, the system should be free of regular interruptions. This type of paper system is cumbersome, requiring personnel to cull, update, and review every card entry made. Agencies currently using paper systems should seriously consider computerized recordkeeping because of its lower long-term cost and ability to process information in seconds instead of hours or days.

Many different computer software programs, known as reference databases, have been developed for this type of recordkeeping. Field information can be manipulated in a tremendous variety of ways, from relational comparisons of different pipe materials and wear, to automatic work order generation based on television inspection done the day before, all generated far faster than written records at far less cost over time. Some of these programs allow all operations (biweekly cleanings, staggered yearly maintenance, or repairs to the system or pumping stations) to be tracked and analyzed over time.

Which software package is most appropriate depends on the size of the system, what the agency wants to accomplish, and what future growth requirements they have. Data management systems can be custom designed or "off the shelf," depending on the capacity of the computer system network currently in place. These programs can be examined at regional and national trade shows or through professional organizations.

Complaints and Service Requests. Whenever a complaint is received from a system user, a service request needs to be recorded on either a paper or a computer-generated form. It is important that these requests be received courteously and investigated promptly. If possible, a repair or maintenance crew can be taken off a less urgent job to correct the trouble. Prompt, effective response and follow-up to complaints pays dividends in the public's attitude toward the agency. If the agency has no nighttime operator on duty, incoming calls outside working hours can be transferred to another 24-hour emergency operation such as the police department. A crew leader on standby, who can be contacted at home, by phone, or by pager, can determine the severity of the problem and call in a designated emergency crew, if necessary.

Service request forms (Figure 3.4) can be easily recorded in either a card file or a computer for use in structuring the preventive maintenance program. These will more readily identify trouble areas in the system. Repairs to these areas can then be planned and budgeted. Computer database systems can effectively help in these service complaints by automatically producing follow-up correspondence to the complainant or generating a follow-up visit order. Most people respond well to this kind of attention, and future problems, if any develop, can be lessened in severity with good follow-up procedures.

Litigation. At some point in time, every agency is faced with a damage claim from an end user because of stoppages. Though not a pleasant experience, it can be one with a successful conclusion if the agency is prepared and if management understands what is involved in a successful defense. For detailed information on litigation matters, the agency's corporation counsel or local League of Municipalities should be contacted to explain the various aspects of the law regarding collection systems operations. The basic tenants of law that apply to collection systems agencies are termed *negligence standards.* Though varying from state to state, the negligence standards focus on four elements.

- Actual notice—defined as notification received from systen. users or field personnel that a problem exists in the system, either a stoppage or a system anomaly. Failure to respond to this notification can lead to a claim based on lack of maintenance, unreasonable time to investigate or resolve, or negligence on the part of the agency towards property.
- Constructive notice—defined as the process of actively looking for system problems on a regular basis. Failure to perform reasonable constructive notice activities can lead to a claim based on lack of maintenance and negligence on the part of the agency.
- Lack of maintenance—defined as the lack of a policy, program, or preparation for maintaining the system to avoid damage to property resultant from stoppages. Failure to regularly, and with reasonable competence, maintain the system can lead to a claim of negligence.
- Unreasonable time—defined as an undue amount of time passing between actual notice and an effective solution of the problem.

The best defense any agency can offer is detailed records derived from a preventive maintenance program. Records showing when a particular line was cleaned or inspected, documentation of proofing efforts, and records of any problems can make the difference between an award or denial of claims. Many agencies have incorporated thorough recordkeeping as a result of litigation and fines or judgments against them. The best way to know what records to keep and how best to handle potential litigation situations is to contact the agency counsel for advice.

Good recordkeeping and prioritized maintenance and repairs are required to keep the system working with minimal disruption. A properly managed system does not have an excessive number of emergency problems. The cost of establishing a preventive maintenance program is money well spent in terms of improving overall system efficiency and community relations and avoiding public and personal property damage and litigation.

NEW REPAIR AND INSTALLATION TECHNOLOGIES

In the last decade, various types of technologies have been developed that allow *in situ* repair and rehabilitation of collection systems. These "trenchless" technologies can allow the installation of new pipeline through underground tunneling or the installation of materials to form a new pipeline wall within an existing but deteriorated line. The objective is to install, replace, or renovate lines with minimum excavation and disturbance. Trenchless repairs are often used in lines under high traffic volume, under existing structures, or in difficult access terrain that does not allow conventional excavation to be done without damage to property or disruption of transportation.

NEW INSTALLATION. Installation of new sewer pipes can be extremely disruptive in areas with a high traffic volume or in densely populated areas. Excavation sometimes is the only solution; however, in many instances, minihorizontal directional drilling or microtunneling techniques can be used with minimal disruption.

Minihorizontal drilling is a surface-launched method for installation of small lines, 0.05 to 0.25 m (2 to 10 in.) in diameter, in lengths of 150 m (500 ft) or more. Typical installation using this technique is approximately 90 m (300 ft), at depths of 5 m (15 ft) or less. Various material pipes can be installed with this process, including polyethylene, PVC, steel, and copper. The drilling process includes the creation of a pilot hole, using mechanical cutting or fluid jetting systems, then enlarging the pilot hole to accommodate the pipe. A slurry typically is used to stabilize the borehole and reduce friction when the pipe is inserted and pulled. The time and space necessary to perform this work are minimal, and recent developments in steering and locating capabilities allow avoidance of obstacles and improve the line and grade of the bore. Soil or geological strata conditions may present some problems in drilling. Soil possessing high gravel content or rock can limit application.

Microtunneling can be used to construct pipelines to very close tolerances in grade and line. Microtunneling consists of a remotely controlled, laser-guided, pipe-jacking system that does not require personnel entry. These systems can be used to install pipelines 450 m (1 500 ft) or longer in a single-pass operation. The pipe can range in diameter from 0.25 m (10 in.) to 3 m (10 ft) and be buried as deep as 30 m (100 ft) below the water table without dewatering. Microtunneling can be used in a broad range of soil conditions ranging from rocks, to boulders, to soft clay, with either a slurry or auger system depending on geological and water table criteria. Pipe materials that can be used with this process include concrete, steel, PVC, fiber-glass reinforced, and vitrified clay.

REPLACEMENT. Pipe replacement, especially in urban areas where multiple underground utilities are present, can create difficulties in surface excavation repairs. Pipe bursting or reaming involves splitting the defective pipe and displacing the fragments to allow insertion of a new pipe of the same diameter. Polyethylene pipe is typically used. Increased capacity can also be obtained with the use of an expanding device, either pneumatic or hydraulic, to provide space to insert a larger line. Upsizing from 0.1-m (4-in.) to 0.2-m (9-in.) pipe has been accomplished in gas, water, and sewer line replacement.

REHABILITATION. Among the many different types of trenchless rehabilitation techniques is the use of cured-in-place pipe (CIPP). This procedure involves the use of resin-activated liners that are inverted or pulled into a deteriorated line and formed and cured using heated water. The water activates the resin, conforming the material to the inside of the pipe walls, and hardens it. After curing, service connections are cut out using a television camera and special cutting tools.

Spot liners are also used for point repairs along sewer lines to rehabilitate short sections of damaged pipe, seal holes, provide a smooth transition past offset joints, and eliminate severe root intrusion problems. Spot liners typically provide structural repairs by overlapping multiple spot liners. Most spot liners are CIPP with fiber-glass–reinforcing mesh. The liners are impregnated with a resin and deployed on a packer or sewer plug, or by a launching device to the repair site. The liner is then expanded and molded to the pipe wall where it is cured by heat application or by the ambient temperature in the sewer. After curing, the deployment equipment is removed and the cured liner is video-monitored.

Another process is the pulled in, fold-in-place pipe, which expands to the internal diameter of the pipe, again using special cutting tools to cut open the service connections. For structural repairs, slip liners can be pulled or pushed into the pipe, effectively placing a pipe within a pipe, making it more structurally sound.

Point repairs can be made in joints using grouting or sealing processes. These processes typically require the use of an inspection camera as a special tool is pulled through the line. When a joint needing point repairs is reached, the tool inflates to seal each end of the pipe, and a sealer or grout is injected into the joint. After it is completed, the tool deflates and is moved to the next site. Grouting and sealing agents can be caustic, and special training for handling, storing, and application should be mandatory before use.

Each of these processes is applicable to a variety of performance criteria, structural characteristics, and flow and volume characteristics. These procedures typically are expensive, though sometimes less so than digging up the existing pipe and disrupting service and roads for long periods of time. When evaluating what form of in-place process to use, it is recommended that an engineer be used to evaluate the structural requirements and other considerations to select the best method for the particular situation.

An additional process for rehabilitation is "sliplining." This process involves pulling a slightly smaller diameter pipe into a defective pipe to improve flow characteristics and eliminate infiltration and exfiltration. Modified "sliplining" uses a liner that is reduced in size to pull into an existing pipe. The liner is then expanded to size to form a tight fit to the pipe walls.

Trenchless techniques are excellent options for repairs, replacement, and renovation. The specific advantages and disadvantages of each technology should be considered for each application. Because this technology is continually developing, not all products and techniques are discussed in depth here. For additional information on these processes, refer to Chapter 6, Construction Contracting.

REFERENCES

American Society of Civil Engineers (1989) *Sulfides in Wastewater Collection and Treatment Systems.* Manual and Report on Engineering Practice No. 69, New York, N.Y.

Water Pollution Control Federation (1979) *Odor Control for Wastewater Facilities.* Manual of Practice No. 22, Washington, D.C.

Water Pollution Control Federation (1987) *Operation and Maintenance of Wastewater Collection Systems.* Manual of Practice No. 7, Washington, D.C.

Water Environment Federation (1990) *Operation of Municipal Wastewater Treatment Plants.* Manual of Practice No. 11, Alexandria, Va.

SUGGESTED READINGS

Frutchey, R., *et al.* (1991). Trenchless Methods in the Replacement and Rehabilitation of Sanitary Sewers. *Oper. Forum,* **8,** 5, 19.

National Association of Sewer Service Companies (1987) *Recommended Specifications for Sewer Rehabilitation.* Maitland, Fla.

National Utility Contractors Association, North American Society for Trenchless Technology, and Water Pollution Control Federation (1991) *North America No-Dig '91 Conference Proceedings.* Kansas City, Mo.

U.S. Environmental Protection Agency (1996) *National Conference on Combined Sewer Overflows.* EPA-625/R-96-007, Washington, D.C.

U.S. Environmental Protection Agency (1993) *Combined Sewer Overflow Control.* EPA-625/R-93-007, Washington, D.C.

U.S. Environmental Protection Agency (1991) *Sewer System Infrastructure Analysis and Rehabilitation.* EPA-625/6-91-030, Cincinnati, Ohio.

Water Environment Federation (1996) *Urban Wet Weather Pollution.* Alexandria, Va.

Water Environment Federation (1995) *Sewers of the Future.* Alexandria, Va.

Chapter 4
Pumping Stations

Wastewater pumping stations, sometimes called lift stations, are used for lifting wastewater to a higher elevation. Pumping stations are installed either in gravity systems, where terrain, groundwater, or other factors can affect the economics of construction, or at the headworks of the wastewater treatment plant.

In addition to representing a major initial capital investment, pumping stations are more expensive to operate and maintain than gravity sewer systems. Resources and maintenance programs must be adequate to protect the capital investment. Special skills are necessary for the operation and maintenance (O&M) of pumping station equipment. Design, construction, and O&M of pumping stations have changed significantly as a result of new regulatory requirements, environmental and liability issues, increasing size, and safety requirements.

Pumping station failure may result in damage to equipment or the environment or endangerment to public health. The goal of pumping station O&M is increased reliability and reduced risk of failure. An effective O&M program is a key element in achieving this goal. Proper design, construction, and operator training are also important.

The management of collection systems and pumping stations has become more critical as technology has changed. This chapter, rather than dealing with specific maintenance tasks for individual pieces of equipment, provides guidelines for a comprehensive pumping station O&M program.

Variations in equipment types, pumping station configuration, and geographical factors determine pumping station design and O&M requirements for different systems. Size also plays a role because large station failures tend to have more catastrophic results. Smaller agencies are disadvantaged because they traditionally do not have the same financial or human resources as larger agencies but are subject to the same environmental and safety regulations.

The information presented in this chapter is intended to aid in the establishment or enhancement of a pumping station maintenance program. It presents proven concepts and techniques that have been developed over time and can be adopted, improved, or modified for an individual operation, regardless of size or location.

Finally, this chapter focuses primarily on the more typical types of pumping stations, rather than those serving systems such as low-pressure vacuum or grinder sewers, which are discussed elsewhere (WPCF, 1986).

INCREASING IMPORTANCE OF PUMPING STATION MANAGEMENT

Careful consideration and maintenance of pumping stations have become more important in recent years for several reasons. Regulatory requirements, for example, have become increasingly stringent and no longer allow bypasses in the event of station failure. These requirements have led to more reliable and sophisticated station designs, such as portable or on-site standby power systems and increased equipment redundancy.

In addition, liability for system failures has increased. Large fines have been levied against agencies and individuals responsible for collection systems when, for example, a station failure has resulted in a violation that could be attributed to poor maintenance programs or negligence.

Formation of regional agencies, which is now common, allows more efficient and cost-effective collection and treatment of wastewater. This has resulted in the installation of larger pumping stations and force mains with higher capacities and operating heads. Pumping stations with design capacities of more than 631 L/s (10 000 gpm) and total dynamic heads greater than

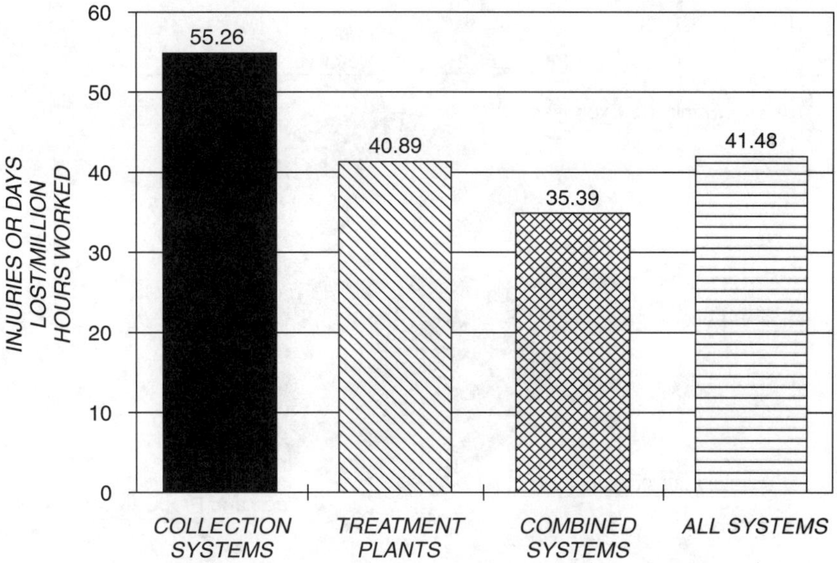

Figure 4.1 Wastewater treatment industry injuries by area (Hadeed, 1990).

60 m (200 ft) are not unusual; nor are force mains longer than 16 km (10 miles). Design techniques and construction methods have changed as a result. Failures (when a bypass or spill occurs) of these larger facilities tend to have a more adverse effect than those of smaller, older facilities.

The safety record for collection systems personnel continues to be poorer than for other segments of either public or private industry. Data from a 1989 safety survey (Hadeed, 1990) indicate that little or no change in this trend has occurred (Figure 4.1) and that many preventable injuries still take place in pumping stations or maintenance shops.

The industry-wide cost of accidents and fatalities, in terms of human and economic resources, is extraordinary. Figure 4.2 illustrates a breakdown of estimated costs of work-related accidents in 1990 (National Safety Council, 1991). Indirect work loss is the value of time lost by uninjured workers. According to the graph, the cost of an accident in terms of work lost by injured workers (including production, handling paperwork, and other administrative details) equals the cost of wages, medical expenses, and insurance administration combined.

IMPROVED TECHNOLOGY. Microprocessor, or computer-based, systems designed for pumping station applications are reliable, cost effective, and functional in many new or rehabilitated pumping stations. Programmable logic controllers (PLC) are used as supervisory pump control systems. Supervisory control and data acquisition systems (SCADA) provide continuous monitoring of the operating status of stations and have the capability to allow station control from a remote location. Instantaneous station alarm reporting,

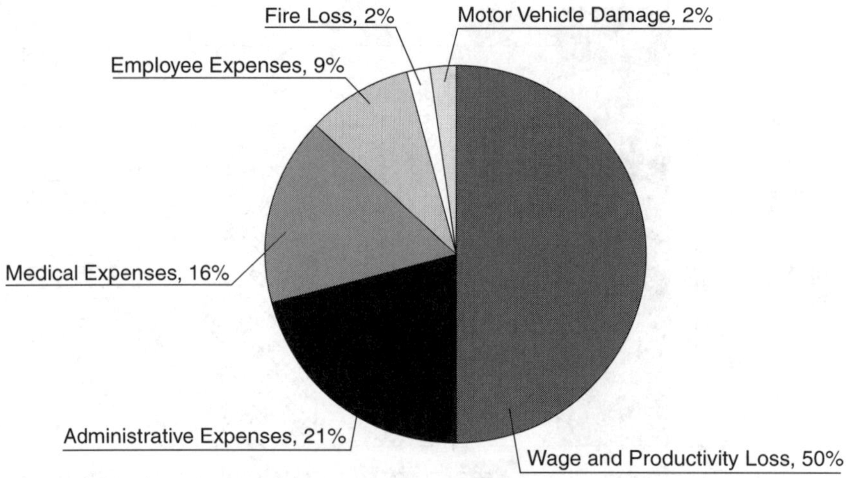

Figure 4.2 Estimated costs of work-related accidents (*Accident Facts* ®, 1998 Edition, Itasca, IL: National Safety Council).

by leased line or radio telemetry systems, allows for faster response times and improved equipment use, and reduces the risk of bypasses. The PLC and SCADA systems can be integrated to provide a remote, real-time pumping station control and monitoring system. Figure 4.3 illustrates a typical application.

Improved design and reliability of variable-speed pump systems provide O&M flexibility and energy savings. Standby or emergency power generators with automatic transfer switches are frequently required, however, during design or retrofitting of pumping stations. These generators represent significant capital expenditures and increased maintenance requirements.

These changes in pumping station design have produced complex systems that have more sophisticated O&M requirements than has traditionally been the case because of their complexity and related environmental and liability considerations. Consequently, those involved with collection systems design, construction, O&M, or management need increased and more refined skills.

PUMPING STATION TYPES

Pumping stations are primarily distinguished by their construction rather than their equipment. Most contain induction motors; centrifugal pumps; electrical distribution and control systems; plug, gate, swing check, knife, and pressure-relief valves; and suction and discharge piping systems. Because of the similarity of the basic equipment, the information in this chapter applies to the following construction variations and equipment:

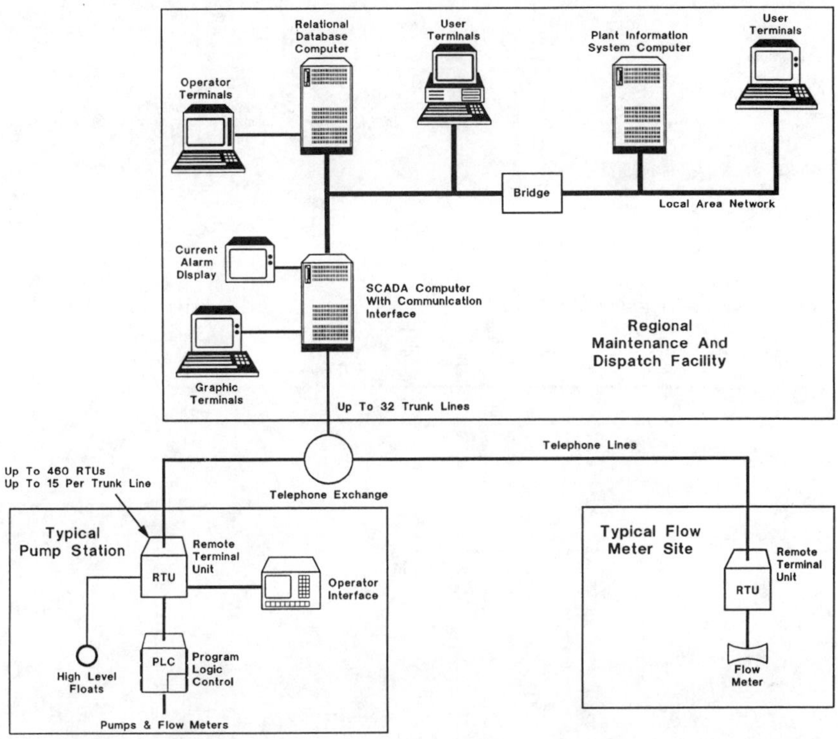

Figure 4.3 Computer-based pump and flow monitoring system.

- Prefabricated, aboveground and belowground;
- Cast or constructed in place; and
- Submersible.

PREFABRICATED PUMPING STATIONS. Prefabricated pumping stations typically are constructed of fiber glass or steel as self-contained units at a manufacturing facility and then shipped to the construction site with all mechanical, electrical, and piping systems installed. Site construction requires the installation of the wet well, which can be prefabricated reinforced concrete barrel sections with suction piping between the dry well and wet well. External force main connections and the necessary electrical, telephone, or other services are also installed on site.

The pumping station may be located above ground (Figure 4.4) or below ground (Figure 4.5). Belowground stations are limited by the number and size of pumps to be accommodated, the station size, and the ability to ship the unit by normal transportation methods to the construction site.

CAST- OR CONSTRUCTED-IN-PLACE PUMPING STATIONS. Cast- or constructed-in-place pumping stations, which are constructed entirely on site, can be either dry-well and wet-well configuration or wet-well only. The latter uses submersible pumps. Cast-in-place stations that use the

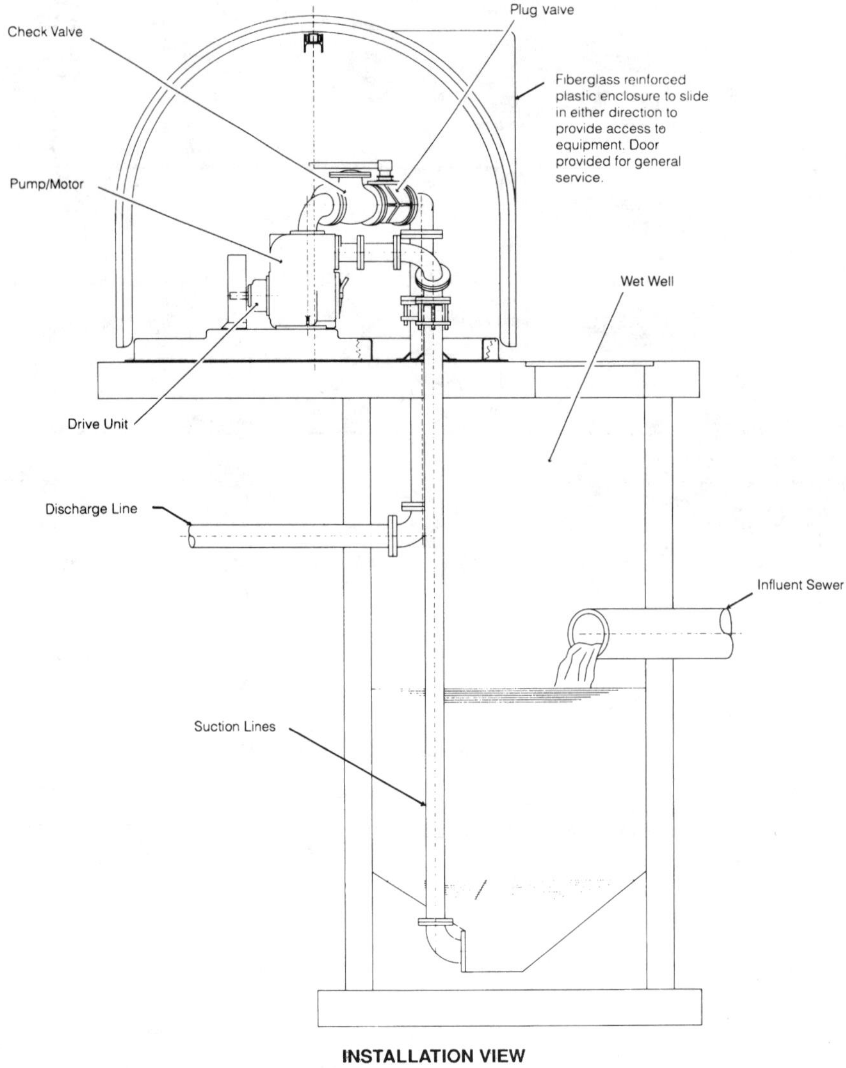

Figure 4.4 Aboveground station.

dry-well and wet-well configuration frequently have a head house containing switchgear, generators, or other auxiliary equipment. This type of construction is more expensive but offers more flexibility in terms of the number and size of pumps and drivers that can be accommodated, wet-well sump considerations, future expansion capabilities, auxiliary equipment, and maintenance. Pumps are installed horizontally or vertically and may be close-coupled, flexibly coupled, or driven through shafting from above. Figure 4.6 illustrates a typical station.

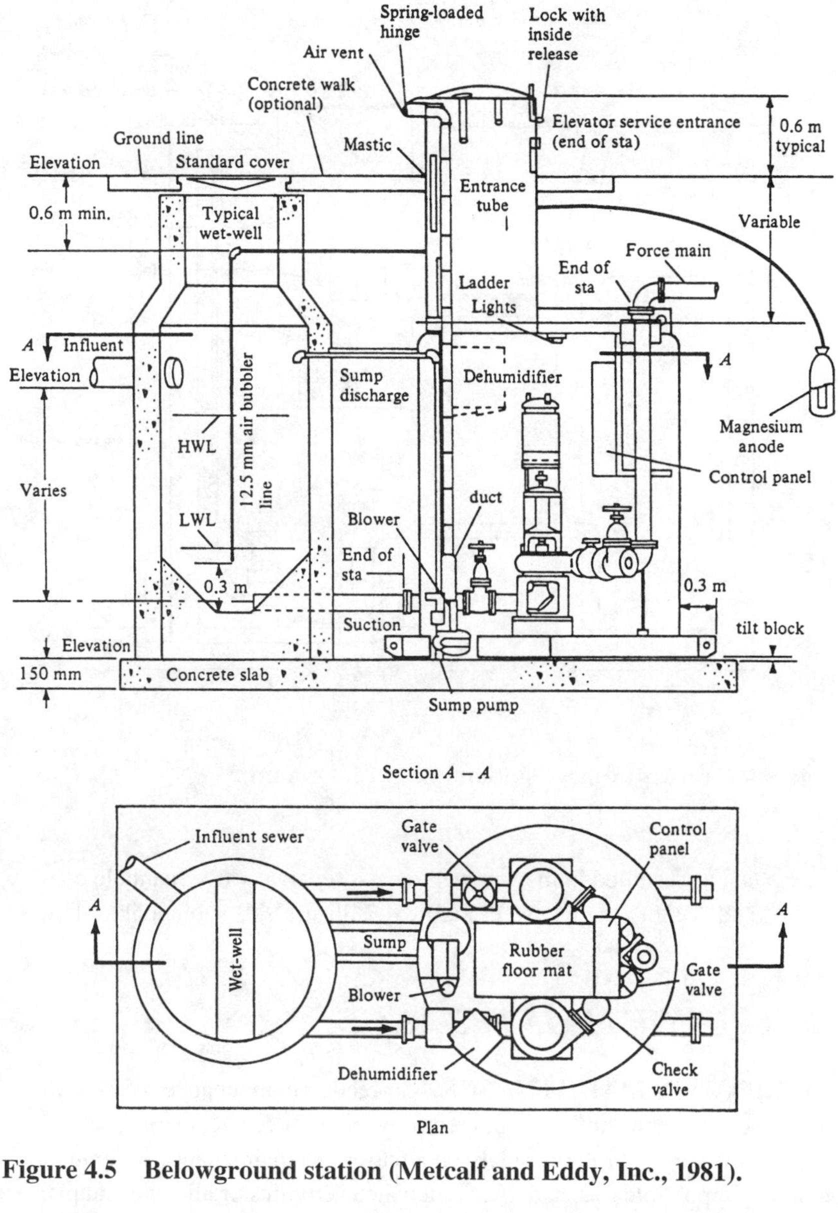

Figure 4.5 Belowground station (Metcalf and Eddy, Inc., 1981).

SUBMERSIBLE PUMPING STATIONS. Submersible pumping stations use cast-in-place or precast concrete sections for wet wells. Submersible close-coupled pumps connect directly to a discharge pipe and are easily installed and removed through fixed rails in the wet well. It is not necessary to enter the wet well to install or remove the pump because it has a self-sealing flange. Electrical controls can be pedestal-mounted in a weatherproof enclosure or a small head house constructed on a slab adjacent to the wet well. To facilitate maintenance and reduce corrosion discharge, check and isolation

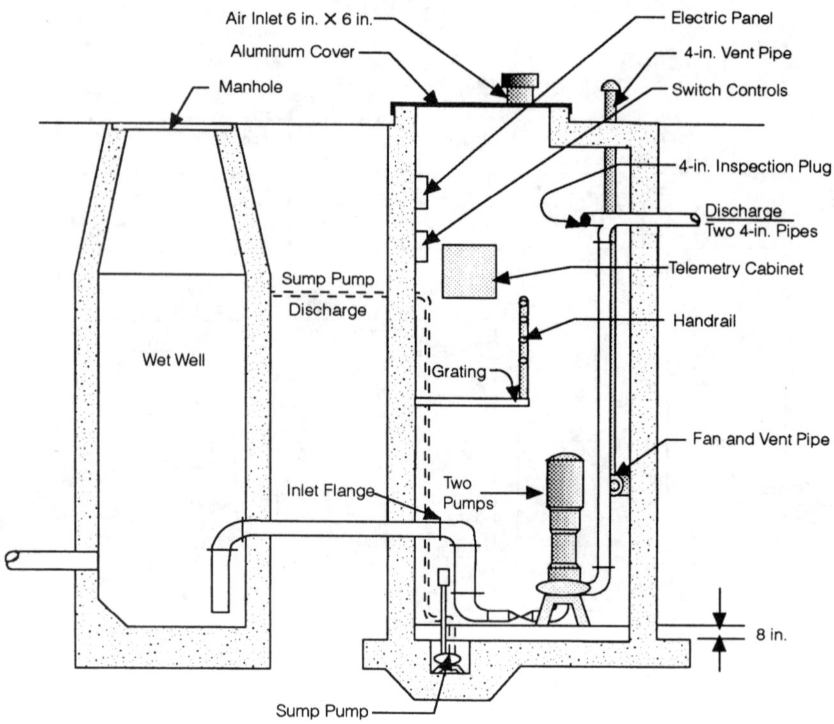

Figure 4.6 Cast-in-place station (in. × 25.4 = mm).

valves can be installed in the wet well or in a separate valve manhole constructed adjacent to the wet well. Figure 4.7 illustrates a submersible station.

TYPES OF MAINTENANCE

CORRECTIVE MAINTENANCE. Corrective maintenance refers to the use of equipment until it fails and requires repair or replacement. The obvious disadvantage of relying solely on a corrective maintenance program is that it is impossible to schedule maintenance activities or allocate and plan budget and staffing resources. Even more importantly, the failure of a critical piece of equipment in the pumping station (for example, the failure of a pump bearing from lack of lubrication) could result in a bypass or flooded basements upstream. Corrective-maintenance-based programs are characterized by

- Unscheduled equipment failures,
- Excessive repair or replacement costs,
- Poor use of human resources,
- Lack of control of expenditures,

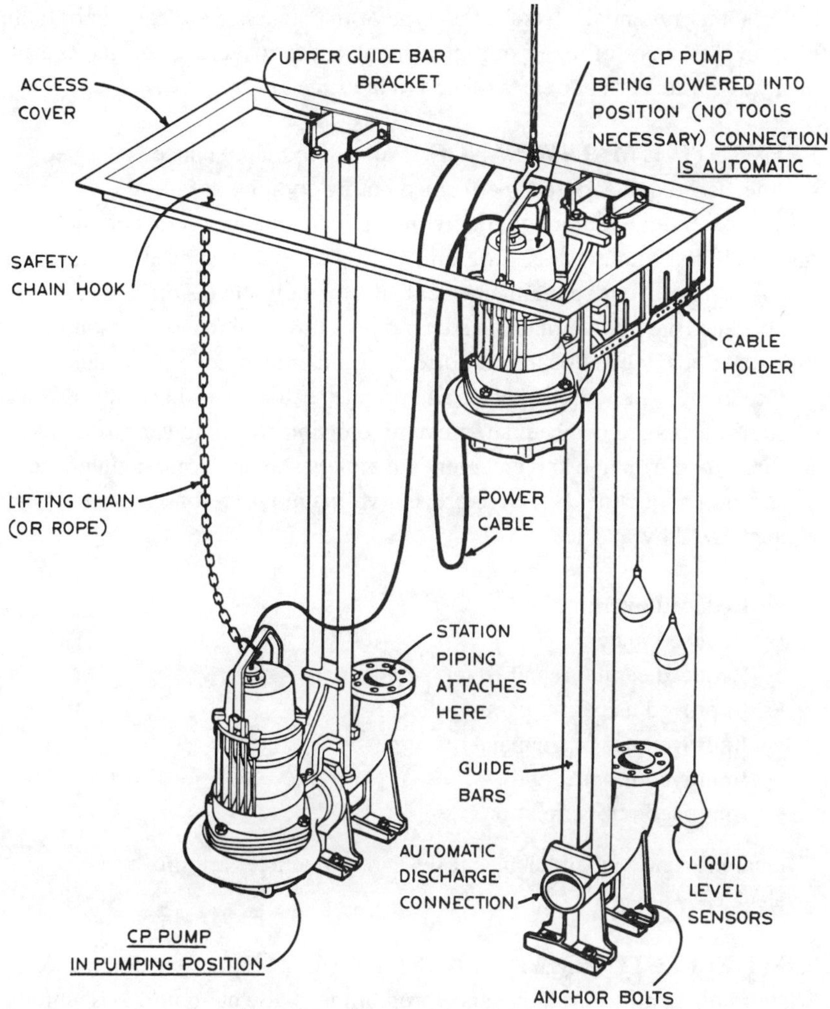

Figure 4.7 Submersible station.

- Poor morale, and
- Increased liability because of negligence.

EMERGENCY MAINTENANCE. Emergency maintenance is a form of corrective maintenance that occurs when a piece of equipment or system fails, creating a threat to public health or equipment. This is a special case of corrective maintenance in which the unscheduled failure of a system or piece of equipment could have a severe or catastrophic effect. Examples of emergencies would be accidental discharges to lakes or streams, backups into homes or businesses, or loss or destruction of equipment.

The cost of emergency maintenance in terms of time and staff resources can be enormous. Furthermore, regulatory agencies frequently view a bypass as a discharge permit violation. Substantial fines and adverse publicity can result. Other potential costs for this type of maintenance are contractor mobilization fees, costs of repair or replacement of destroyed or damaged equipment, overtime costs, or costs related to lawsuits.

PREVENTIVE MAINTENANCE. Preventive maintenance is defined as routine, scheduled activities performed before equipment failure for the purpose of extending equipment life, reducing maintenance costs, and increasing reliability. Maintenance programs should devote an appropriate level of resources to preventive maintenance. Unfortunately, this is difficult in practice because preventive maintenance requires time and material resources that may already be allocated to corrective and emergency maintenance.

Experiences in both the public and private sector show, however, that as more resources are dedicated to preventive maintenance, fewer resources are required for corrective and emergency maintenance, and total maintenance costs begin to decline. Effective preventive-maintenance-based programs are characterized by

- Reduced overtime costs,
- Reduced material costs,
- Reduced capital repair or replacement costs,
- Improved use of human resources,
- Improved use of financial resources,
- Improved morale, and
- Improved public relations.

Typical preventive maintenance tasks and frequencies are shown in Table 4.1.

COST-EFFECTIVE MAINTENANCE. An important question to ask when implementing a maintenance program is, "At what point does a preventive maintenance program cease to be cost effective?" This question, unfortunately, is difficult if not impossible to answer quantitatively because the effect of a poor preventive maintenance program on emergency maintenance costs is unpredictable. Experience does show, however, that the initial expense preventive maintenance incurs typically is repaid in longer and more trouble-free operation. The following example may illustrate the benefits of preventive maintenance versus the unpredictable cost of emergency maintenance.

A pumping station fails because of a lack of maintenance. The failure results in a bypass to an environmentally sensitive stream. The failure results in emergency expenditures not only to restore operation but also to cover cleanup costs and the resulting fine levied for a permit violation. These costs could easily far exceed the entire annual O&M budget for the station. Managing pumping station maintenance programs today requires the consideration of other criteria in conjunction with budgetary factors to determine the level of maintenance to be performed. Another possible approach is risk

Table 4.1 Preventive maintenance tasks and frequency.[a]

System	Daily	Weekly	Monthly	Quarterly	Semi-annually	Annually	As needed
Electrical							
Medium-voltage switchgear				1[a]		2	
Phase transformer(s)				1		2	
Incoming service				1		2	
Main circuit breaker/disconnect				1		2	
Metering section				1		2	
Ground fault						2	
Automatic transfer switch		1					
Bus, vertical and horizontal						2	
Bus cables/stab-ins						2	
Motor control center			1		2		
Branch circuit breakers				1		2	
Motor starters		1		2			
Motor overload relays					2		
Relays						2	
Indicating lights	1		2				
Voltage/current readings	1						
Infrared scan						2	
Power factor capacitors		1		2			
Supervisory control							
Relays				2			
Wiring			1		2		
Fuses		1		2			
Circuit breakers					2		
Pump sequencing		1			2		

(Continued)

Table 4.1 Preventive maintenance tasks and frequency[a] (continued).

System	Daily	Weekly	Monthly	Quarterly	Semi-annually	Annually	As needed
Pump operating levels		1			2		
Compressors, bubbler		1				2	
Pneumatic piping/valves				1		2	
Pressure switches		1			2		
Alarms and levels	1				2		
Indicating lights	1						
Floats	1					2	
Pumps							
Bearings (vibration)		1				2	
Alignment				1		2	
Coupling				1		2	
Lubrication						1	1
Packing						2	
Mechanical seals		1				2	
Mounting						2	
Balance						2	
Impeller clearance						2	
Capacity		1					
Suction pressure		1					
Discharge pressure		1					
Station calibration						1, 2	
Running time	1						
Overhaul							2
Motors							
Bearings			1			2	

(Continued)

Table 4.1 Preventive maintenance tasks and frequency[a] (continued).

System	Daily	Weekly	Monthly	Quarterly	Semi-annually	Annually	As needed
Balance						2	
Conduit box connections						2	
Lubrication							1
Insulation resistance						2	
Running temperatures		1				2	
Full load amperes		1					
Voltage		1					
Vibration analysis						2	
Alignment		1				2	
Coupling		1				2	
Overhaul							2
Valves							
Suction				1		2	
Discharge				1		2	
Check		1				2	
Pressure relief		1				2	
Gate				1		2	
Plug					1		
Miscellaneous							
Piping						1	
Pipe restraints				1		2	
Bubbler system	1			2			
Auxiliary systems							1,2

(Continued)

Table 4.1 Preventive maintenance tasks and frequency[a] (continued).

System	Daily	Weekly	Monthly	Quarterly	Semi-annually	Annually	As needed
Water	1			2			
Sump	1			2			
Blowers/fans	1			2			
Heating system	1			2			
Cooling system	1			2			
Engine generator sets	1		2				
Lighting	1			2			

[a] Level 1 maintenance is performed by the operator. Level 2 maintenance is performed by maintenance personnel.

management, or determining what maintenance has to be done to reduce the risk of failure to a reasonable level. Examples of other criteria that typically are considered are equipment reliability, peak-flow capacity, effects of a station failure, frequency and magnitude of failures, and design deficiencies.

PREDICTIVE MAINTENANCE. Predictive-maintenance-based programs are being increasingly used in the public sector. This increase has resulted from developments in the instruments required for this type of system, which have become more portable and less expensive. Predictive maintenance is well suited to pumping station maintenance. Predictive maintenance is a form of preventive maintenance that measures the condition of a piece of equipment or machine and then predicts the time remaining before failure of the machine or a component occurs. As a result, corrective maintenance can be scheduled on a more rational basis and at less cost. A variety of diagnostic techniques is used to assess equipment. Examples are discussed in the following sections.

Vibration Analysis. Computer-based vibration analysis requires the use of microprocessor-based portable field instruments to measure the frequency versus amplitude of equipment vibration. The information can be analyzed on site and downloaded to a personal computer (PC) to establish and store baseline vibration levels versus frequency. This method allows the detection of equipment problems earlier than is possible with other indicators such as heat or audible noise.

Vibration analysis can detect a variety of electrical and mechanical problems, including

- Rotation unbalance;
- Bent shafts;
- Defects in bearings, inner or outer races, and balls or rollers;
- Mechanical loosening; and
- Motor electrical and magnetic anomalies.

Baseline information ideally is obtained as part of the acceptance process for new pumping stations or when the equipment has been overhauled. Periodic measurements compare the vibration levels with the original baseline data and, based on the rate of change, predict failure. Vibration levels can also be compared to industry standards that indicate equipment condition as a function of type or class and operating speed.

Infrared Scanning. Another effective method of predictive maintenance is the use of scanning infrared radiation to detect temperature changes. Small portable cameras, designed to measure radiation levels and convert these to temperature, are used to scan electrical and mechanical equipment and identify "hot spots." This process is particularly useful for locating potential problems in electrical systems. Infrared scanning typically is used for transformers, main disconnects, branch circuit breakers, motor starters and overload relays, conductors, motors and motor connections, fuses and fuse holders, and capacitors.

Predictive maintenance is a valuable and cost-effective tool that can be applied for stations where reliability is critical and the consequences of an equipment or station failure could be significant. It is not necessary for an agency to buy the diagnostic equipment and train in-house maintenance staff. There are specialized contractors that provide these services.

An effective preventive maintenance program results in the highest reliability and availability of a pumping station. However, even the best preventive maintenance program will not eliminate the need for corrective or emergency maintenance efforts because it is impossible to design a system to account for natural disasters such as hurricanes, earthquakes, torrential rainfalls, or other emergencies. Thus, a good overall maintenance program encompasses elements of all three types of programs.

MAINTENANCE ORGANIZATION

The organization and staffing required to support a pumping station maintenance program depend on the following types of systems in the station:

- Electrical and electronic systems,
- Mechanical systems, and
- Hydraulic and pneumatic systems.

Operation and maintenance for each of these systems requires a different set of skills and varying levels of knowledge. Because it would be extremely unlikely for any single employee to have the detailed knowledge required to operate and maintain all of these systems, it is typical for an agency to separate O&M employee classifications.

Operating responsibility, which would encompass routine maintenance of the pumping station, is assigned to one group of employees. System and equipment maintenance is assigned to a second group, whose members have specialized skills in electronic, mechanical, hydraulic, and pneumatic systems. The second group might include pipe fitters, electricians, and machinists. Typically, the tasks to be performed and the skills required to perform the tasks form the basis of the classifications. It is not necessary to have a separate classification for each craft if the organizational structure addresses what work is to be done, what skills are necessary to do the work, and who is going to do it. Figure 4.8 illustrates a typical example of O&M staffing.

Many state and local codes require formal education, certification, or licensing to work on specific systems. Because of safety issues, maintenance of electrical, low- and high-pressure hot water, or steam boiler systems may require a state or local license.

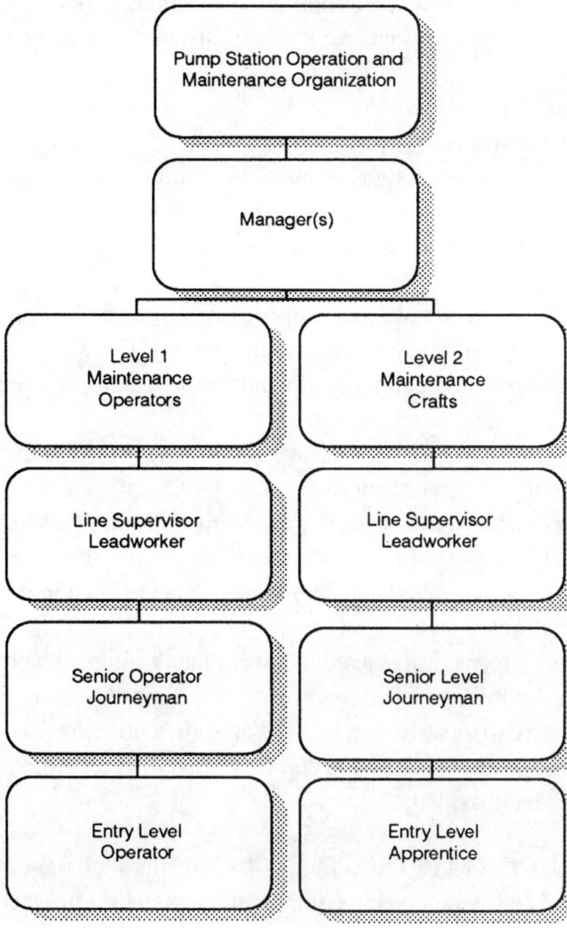

Figure 4.8 Hierarchy of station personnel.

OPERATION STAFFING. Typical operator responsibilities may include

- Performing routine pumping station inspections;
- Recording or interpreting information from elapsed time meters;
- Alternating pumps;
- Performing station housekeeping;
- Cleaning bar screens and wet wells;
- Verifying flow information;
- Exercising emergency standby power systems;
- Adjusting packing glands;
- Testing alarm systems;
- Checking parameters such as incoming voltage and current using hard-wired meters;
- Performing equipment preventive maintenance such as lubrication, belt tightening, checking of fluid levels, and exercising of equipment on a regular schedule;

- Responding to alarm conditions at the pumping station;
- Diagnosing problems and requesting assistance to resolve them when necessary;
- Initiating work orders;
- Assisting craft maintenance crews;
- Providing input on design or station rehabilitation;
- Participating in the budgeting process to identify resources; and
- Supervising field operations.

There may be from one to four levels of experience in the operator classification—entry level operator, journeyman operator, line supervisor/lead operator, and manager—depending on the number and size of stations in the system.

Additional responsibilities for these classes frequently include pipeline inspection, cleaning, and maintenance.

Each of these classes has specific job descriptions that include the tasks involved, skills and level of education required, and performance criteria. In establishing a program, many agencies have used certification criteria to define job descriptions.

Certification information is available from many sources, including the Association of Boards of Certification program guides and state certification agencies. Before certification can be used as an incentive for job advancement, it must be strongly supported and encouraged, and sufficient financial resources must be provided.

MAINTENANCE STAFFING. Figure 4.8 illustrates a hierarchy of craft or trade maintenance for supporting equipment maintenance of electrical, mechanical, and hydraulic/pneumatic systems: manager, lead person/line supervisor, and journeyman. A number of agencies also use a fourth class, apprentice journeyman, administered through an established formal apprenticeship program. Depending on the size, structure, and organization of the agency, the managers for each of the two levels of O&M can be individual managers, one assigned to maintenance and the other assigned to operation, or a single manager responsible for both operation and maintenance.

Typically, craft personnel are responsible for performing the following:

- Low- and medium-voltage motor control center maintenance,
- Supervisory systems maintenance,
- Instrument and control system maintenance,
- Internal motor maintenance,
- Lighting panel and branch circuit maintenance,
- Pump overhaul,
- Mechanical seal replacement,
- Vibration measurement analysis,
- Pneumatic systems maintenance,
- Valve maintenance and overhaul,

- Engine repair and overhaul,
- Generator maintenance, and
- Heating, ventilation, and air conditioning system maintenance.

Although each member of the O&M crew has specific skills and task assignments based on skills and jurisdictional issues, a successful maintenance program integrates and uses these skills to create a team. For example, the removal of a pump from a pumping station for overhaul may require an electrician to disconnect motor connections, an operator to reset the supervisory control for pump sequencing, and a mechanic for rigging and removing the pump.

Approaching pumping station maintenance as a team assignment facilitates efficient resource use. One effective approach is to develop composite crews on a skills-required basis and use crew members within their jurisdictions. In the preceding example, the electrician and mechanic would be an established crew, assigned work that requires both sets of skills.

Cross training is another way to increase maintenance effectiveness. Maintenance personnel are often unable to properly diagnose the cause of a pumping station system failure because pumping station systems tend to be specialized. Premature mechanical seal failure in a pump could be the result of excessive vibration in the motor, pump, or drive shaft or a critical frequency in the entire system. Nonspecialized maintenance personnel may have difficulty identifying the source of the problem or may at least take an inordinate amount of time to do so.

ESTABLISHING A MAINTENANCE PROGRAM

LEVEL ONE MAINTENANCE. Level one maintenance requirements are completed by operators who visit the station on a periodic basis, daily to weekly, to perform a variety of preventive maintenance tasks. The frequency of the visits is a function of the size of station, type of station, level to which the maintenance is critical, reliability of the station, crew availability, and remote monitoring or supervisory control capability.

Level one maintenance is effected by operators who are responsible for observing and verifying the operation of all systems in the station; performing maintenance that does not require access to the electrical system or the dismantling and repair of equipment that requires specialized skills; and ensuring overall station reliability and availability. The operator still needs a broad background and understanding of the multitude of equipment and systems in a station. Operators play a major role in maintaining the reliability of the station.

Typical tasks performed by the operator include observing, recording, and analyzing incoming line voltage and current; pump operating levels, times, and discharge pressures; station flows; or other abnormal conditions.

Qualified operators interpret and use this information to troubleshoot both internal and external station problems. Historical information is useful for identifying deviations from baseline conditions and the potential causes. For example, increased dry-weather flow is frequently caused by new upstream sources of inflow and infiltration. A computer database can facilitate data analysis and monitor trends and changes. Indicators of system problems might be

- Utility-supply problems such as low or high supply voltage or unbalanced three-phase power or pumping station changes in load, grounding, or cabling;
- Float or bubbler pipe problems in wet wells, pneumatic system problems, or changes in the supervisory control system;
- Pump performance or supervisory control problems, or increased or decreased station flow;
- Internal versus external station problems or isolation of external problems; or
- Force main problems, suction condition changes, or poor pump performance.

LEVEL TWO MAINTENANCE. Level two maintenance typically is performed on a less frequent basis but requires specialized skills in specific areas and equipment systems. This level typically is provided by personnel with the skills equivalent to journeyman electricians, pipe fitters, machinists, and mechanics.

Composite crews assembled to meet the needs of a specific task are the most effective. A crew might be composed, for example, of an electrician and pipe fitter or machinist, depending on the work assignments. Composite crews can be difficult to implement if crews have been structured differently in the past and jurisdictional issues have to be resolved. These crews also work closely with the operators in all phases of maintenance.

FREQUENCY OF MAINTENANCE. The prescribed frequency of level one and level two maintenance activities depends on many variables. As a result, no single schedule meets the needs of every maintenance program. Table 4.1 illustrates some of the various equipment systems in a pumping station and typical frequencies for level one and level two maintenance. These are only guidelines and can be modified according to the following variables:

- Station capacity;
- Previous history;
- Reliability;
- Size;
- Operating environment, including temperature and humidity;
- Running time;
- Deviations from manufacturer-recommended operating parameters;

- Level to which the station is critical; and
- Seasonal considerations.

EQUIPMENT AND TOOL REQUIREMENTS. Pickup trucks with utility boxes that are rated at 1 t (1 ton) or higher load capacity are typical vehicles for operators. Four-wheel-drive vehicles typically are used to accommodate different weather and access conditions. Confined space and traffic control equipment, safety equipment such as first-aid kits and gloves, a hoist, hand tools, and preventive maintenance materials and supplies are carried on the truck. These vehicles can also be outfitted for other applications, for example, by providing snow-plowing equipment or standby generators and pumps.

Vehicles used for second-level maintenance are more specific and can range from a utility-type pickup to van-type vehicles. In some cases, they may be set up as a "shop on wheels," outfitted for major overhauls at the pumping station with safety and rigging equipment, hand and electrical tools, diagnostic instruments, mechanical and electrical material and supplies, and welding and cutting equipment.

EMERGENCY RESPONSE. Even where excellent pumping station maintenance programs are in place, a station failure can occur at any time (for example, as a result of a natural disaster). Toxic spills or discharges can also create station operating problems. Emergency equipment and systems are included as part of the pumping station maintenance program. A risk analysis is performed for each pumping station to define the possibility and effects of a prolonged station failure. Response measures are then developed. The equipment required is identified and a long-range plan is developed to acquire the equipment and develop response programs. Each station analysis includes, at a minimum,

- Peak flow storage,
- Lowest upstream manhole or basement,
- Peak flow rate,
- Bypass arrangement,
- Equipment necessary for bypass,
- Emergency power requirements,
- Utility emergency phone numbers, and
- Utility account and circuit numbers.

Local contractors are a resource in an emergency. Each year, electrical, mechanical, and general contractors in the area can be formally or informally surveyed for the equipment available, specialties, labor rates, equipment rates, and capabilities.

In many cases, local or state law dictates that formal procedures be followed when expenditures exceed certain levels. The emergency response plan addresses this by authorizing a senior manager to declare an emergency and mobilize the resources necessary to reduce the effects of the emergency.

In cases where equipment is not available, the agency must purchase it. Examples of standard equipment are

- Portable engine-driven generator sets (30, 85, 150, and 300 kW [40, 115, 200, and 400 hp]). The size and weight of the trailer and generator set may not allow portability without specialized towing equipment if generators have capacities higher than 300 kW.
- Hydraulic or electric submersible pumps (10, 15, or 25 cm [4, 6, or 10 in.]).
- Flexible hoses (10, 15, 20, and 25 cm [4, 6, 8, and 10 in.]) and 25-cm (10-in.) aluminum irrigation pipe.
- Quick-disconnect fittings.
- Check and isolation valves.
- Motor controls for electric pumps.
- Level control system.

EQUIPMENT REPAIR. Pumping station equipment may be repaired in one of three ways: contract repair, in-house repair, or a combination of both. Contract repair typically is used by small agencies that do not have their own second-level maintenance personnel. Most larger agencies use a combination of in-house and contracted maintenance, also based on size and economics. For example, few agencies have in-house motor rewind capabilities.

If in-house repair is used, the appropriate equipment must be available. This should reduce the occurrence of accidents, such as back injuries, that are caused by inadequate material handling equipment like hoists and cranes. Productivity is also increased and damage to pumping station equipment is reduced if the right tools are available and used.

Size, capacities, and other specifications of shop equipment (for example, the maximum impeller diameter that a lathe can accommodate when trimming an impeller) are determined by the pumping station equipment in a given system.

Typical shop equipment would include a lathe, mill, bank saw, metal saw, sand blasting booth, bridge crane, drill press, brake, shear, arbor press, welding equipment, grinder, press, balancing machine, and electrical test benches.

WAREHOUSE AND INVENTORY. A well-organized and stocked warehouse is integral to any pumping station maintenance program. At least three factors are considered when stocking material: the degree to which the part is critical, the lead time from the manufacturer, and the frequency of use. In many cases, mechanical parts can be fabricated in the shop or purchased locally at less cost. Table 4.2 lists typical equipment kept in a warehouse.

COMPUTER-BASED SYSTEMS. An equipment-oriented maintenance management system can reduce the time and cost of pumping station man-

Table 4.2 Equipment kept in a warehouse.

Electrical
Circuit breakers and fuses
Starters and overload relays
Instrumentation replacement parts
Control relays
Floats
Solid-state control system parts
Consumable items
Fractional horsepower motors
Wire and cabling
Hydraulic and mechanical
Mechanical seals and packing
Bearings
Power frames
Impellers
Grease shields
Shaft sleeves
Wear rings
Valve shafts, disks, and hardware
Vehicle parts
Bubbler compressors
Pipes and hoses
Valves
Heating, ventilation, and air conditioning parts

agement. Computer-based maintenance management systems offer additional advantages. To appreciate the amount of information that must be stored, tracked, and integrated, one only need analyze the steps and information required to order a replacement pump part:

- Issuing a work order,
- Obtaining equipment nameplate and master parts list data,
- Checking spare parts inventory, and
- Purchasing or replacing the part in inventory.

This can be a time-consuming task when performed manually. For computer-based systems, information can be stored in databases and retrieved quickly to expedite the process. Many approaches can be taken with this type of system, ranging from simple databases designed in house to networked systems that run on mainframe computers. Numerous "off-the-shelf" systems are also available that operate on desktop PCs or workstations. Functions for these systems include

- Storing all relevant equipment nameplate data,
- Generating work orders,
- Tracking inventory and warehousing operations,

- Tracking maintenance history,
- Generating purchase orders,
- Keeping records of maintenance activity, and
- Reporting.

Computer-based systems depend on data input from existing manual systems or databases developed at the time the computer-based system is installed. The computer will be of no use if data are not available. The advantage of computer-based systems is the ability to store, retrieve, and manipulate data quickly.

Recordkeeping is another function that PCs perform well. Documentation of training, for example, is extremely important if a station or system failure results in litigation. Computer-based budgeting and accounting functions are necessary to optimize budget control and analysis and evaluate maintenance program performance. Analysis of flow data, rainfall, pump running times, and power consumption are all done on a routine basis to assist in data management.

Although computer systems offer tremendous potential for improving maintenance and operational management of pumping stations, many agencies have not yet recognized their value. In many cases, computer systems are underused because of a disinterest or lack of resources. Resource commitment and the involvement of staff at each level facilitate success.

DESIGN. Design deficiencies continue to cause problems in pumping station O&M. As noted in Chapter 2, design problems can be reduced through the transfer of information between the designers and experienced O&M staff and the development and implementation of standard specifications for pumping stations. Examples of design problems are shown in Table 4.3.

Pumping station O&M personnel are involved in every station design at the project planning stage, including 40% design review, 70% design review, and 90% design review.

Standard specifications for pumping station design include

- Equipment performance standards (manufacture, acceptance);
- Detailed construction standards, materials, and methods;
- Start-up process;
- Acceptance criteria; and
- Detailed O&M manuals.

BUDGETING FOR PUMPING STATION MAINTENANCE. Figure 4.9 shows an operating budget, excluding direct and indirect labor, for a network of 60 pumping stations. Of note is the amount of the budget that is "fixed" from year to year except for inflation and increases or decreases that would affect fixed costs. This budget is for pumping station maintenance only and does not include gravity sewer cleaning or maintenance.

Table 4.3 Typical design problems.

Wet-well	
	Inadequate access for cleaning/maintenance
	Entrained air
	Inadequate pump submergence
	Wrong material selection (ladders, grates, fasteners, ventilation)
	Electrical equipment selection/installation
Pump	
	Wrong pump selection (based on wrong hydraulic system analysis)
	Inadequate net positive suction head (NPSH)
	Excessive vibration (critical frequencies, base-plate installation, pipe strain, imbalance, misalignment)
	Performance problems (manufacturer-caused)
Motor	
	Wrong selection (code, service factor, insulation)
	Overload problems
	Vibration
	Excessive starts per hour

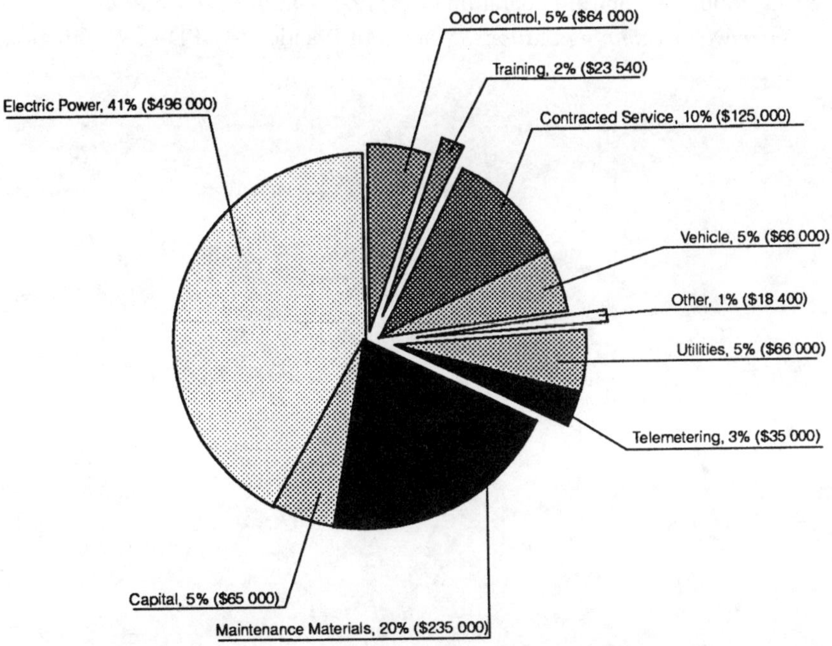

Figure 4.9 Breakdown of pumping station maintenance budget for the Metropolitan Waste Control Commission, Twin Cities metropolitan area. Data do not reflect annual salary and benefit expenses of $1.2 million and $ 293 000, respectively.

*R*EFERENCES

Hadeed, S.J. (1990) 1989 Safety Survey: Injury Rates Drop to 15-Year Low. *Oper. Forum*, **7**, 4, 24.

Metcalf and Eddy, Inc. (1981) *Wastewater Engineering*. McGraw-Hill, Inc., New York, N.Y.

National Safety Council (1998) *Accident Facts ®,* Itasca, Ill., 1998 Ed.

Water Pollution Control Federation (1986) *Alternative Sewer Systems*. Manual of Practice FD-12, Washington, D.C.

*S*UGGESTED READINGS

Karassik, I., *et al.* (1976) *Pump Handbook*. 2nd Edition, McGraw-Hill, Inc., Highstown, N.J.

Metcalf & Eddy, Inc. (1981) *Wastewater Engineering: Collection and Pumping of Wastewater*. G. Tchobanoglous (Ed.), McGraw-Hill, Inc., New York, N.Y.

Water Pollution Control Federation (1989) *Emergency Planning for Municipal Wastewater Facilities*. Manual of Practice SM-8, Washington, D.C.

Water Pollution Control Federation (1981) *Design of Wastewater and Stormwater Pumping Stations*. Manual of Practice No. FD-4, Washington, D.C.

Chapter 5
Government and Community Relations

Collection systems personnel and activities are the most visible segment of any wastewater agency. On city and town streets, telephone, electric, water, and gas company workers project a public image for their respective utilities. This chapter suggests ways in which collection systems employees can enhance the image of their agencies.

PERSONAL RESPONSIBILITIES

The first rule of government and community relations is "know thyself." Effective managers of collection systems understand their function within the entire wastewater agency. They are aware of their responsibilities in maintaining a wastewater collection system and can articulate these responsibilities. They are familiar with the scope of services that they provide. The agency's responsibility may only include certain interceptor sewers or trunk lines, or it may include pumping stations, metering stations, lateral lines, force mains, and other facilities and structures. If another agency is responsible for tributary facilities within the system, managers should be aware of this and be able to supply pertinent information.

Because most of the collection system is within nonagency private or public rights-of-way and easements, collection systems managers should be familiar with the areas around these locations and the legality of their right to gain access to service these facilities. They should know the legislation or ordinances that created the agency and where a citizen can get a copy of the legislation.

Managers are accountable for their immediate responsibilities and should also know who to contact concerning facilities outside their responsibility.

PUBLIC RESPONSIBILITIES

Public collection systems managers have to be aware of other utilities, including gas, water, electric, telephone, and storm sewer systems, that share public rights-of-way and private easements. Managers keep records of the names, telephone numbers, and contact people for other utilities. The manager and supervisors within a collection system should also be familiar with the local numbers used to notify other utilities of an expected roadway excavation and request that they identify the location of their respective lines within a right-of-way. Participation in an area "dig safe" program is a must to protect sewer lines from other excavation.

Familiarity with other utilities is also helpful in answering citizen inquiries. If a roadway problem falls outside the responsibility of the collection system, a supervisor or employee within the collection system can contact other utilities or provide an interested party with the appropriate contact person and telephone number.

Knowledge of other state and local municipal agencies is useful because they may work closely with collection systems personnel when maintenance or repair of a specific facility requires disruption of a public roadway. The system manager may have to attend public meetings concerning the expected disruption. The public works department guides the manager of the project in reducing public inconvenience and ensuring public safety. Local police, fire, and emergency medical services are contacted when traffic disruption is expected.

Managers should communicate their wastewater agency's hazardous contingency plan or policy to all collection system employees, especially supervisors. These plans may involve local emergency agencies such as fire, police, or rescue departments or hospitals.

RELATIONSHIP WITH GOVERNMENT AGENCIES

State environmental agencies typically administer National Pollutant Discharge Elimination System permits to regulate the wastewater treatment system's discharge. The state agency may levy penalties or fines for unpermitted or noncompliant discharges. The magnitude of these penalties may depend on the level of cooperation offered by the agency. Collection systems managers work closely with the government agency and supply all records, reports, and information pertaining to the discharge.

A more personal relationship should be established with other collection systems managers or public works personnel in the region. This relationship could be invaluable in giving advice, sharing information, or providing assistance in solving problems.

The large wastewater agency of which a collection system is a part may have relationships with government agencies not contacted by the collection systems manager. It would then fall to the wastewater agency to promote the services of the agency. The collection systems manager can assist in this promotion or marketing of sewer services if the agency so desires.

EMPLOYEES AS AMBASSADORS (PUBLIC RELATIONS)

Employees of government entities, municipal agencies, and collection systems frequently contact each other on a management level. Community members and contractors, however, typically are in direct contact with collection systems employees. Communication with employees is the only contact most people have with the wastewater agency. It is important that these employees create an image that reflects positively on the agency. Supervisory and clerical employees responsible for answering telephones, for instance, should be trained in customer service.

The external appearance of collection systems crews in the field influences the public's confidence in the agency. The agency should provide uniforms for its crew members and ensure that all equipment is well maintained. Projecting a positive professional image also helps improve employee morale.

How employees handle equipment influences public perception of the agency.

Proper and courteous driving practices are emphasized (for example, employees driving agency vehicles should always extend the right-of-way to the general public). Trucks and vehicles are marked with decals that identify the agency.

Employees who behave in a courteous, polite, and helpful manner leave a good impression with the public and create a positive image for the agency. This image helps in the future when the agency has to approach the public or government for rate or salary increases, purchases, and other resources.

Sometimes the public may not understand the importance of maintenance or repair operations that disrupt their routine activities. Helpful, courteous, and understanding employees create positive first impressions and promote smoother relations between the agency and the public.

PUBLIC INFORMATION PROGRAMS

Collection systems managers provide a vital service to all citizens within the community. Wastewater professionals are aware of the significance of this service, but the community typically knows little about the industry. A responsible agency educates the public about the industry. Because the community may only be aware of the agency through articles in local newspapers that do not always create a positive image, the agency must promote itself and its collection system.

Agency representatives can make speeches and presentations to local schools about the wastewater field. Science teachers are encouraged to promote student education regarding the wastewater industry.

The agency can also participate in local or statewide community activities, such as fairs, historic events, and craft shows, which promote local activities. For example, a booth staffed by agency employees might highlight recent projects, with photographs or a television monitor showing a video inspection of the interior of large sewers. Such a display is particularly effective when combined with pictures of local rivers and harbors while emphasizing the role of collection systems in preserving waterways. A visual presentation of the agency's activities at the treatment plant with charts and graphs also communicates the agency's goals of maintaining the quality of nearby waterways.

Most communities and states have local environmental groups. These groups act as "watchdogs" in overseeing the activities of any public agency or private company whose activities can adversely affect receiving waters. It is important for the agency and the collection system to become aware of these groups and work with them. Attendance at their public meetings, educational workshops, or award ceremonies helps convey the message that the agency and the environmental group are working toward common goals.

Professional organizations offer publications and audio and visual aids on the wastewater industry and environmental concerns. This information can be made available to the public through handouts when the opportunity presents itself.

If the agency feels a need to impart important information such as illegal connections, sewer charges, or functions of sewers, this information can be included when sending out individual tax or sewer bills. Newer public access outlets such as cable television and the Internet can be used by the agency to communicate with the community.

RELATIONSHIP WITH CONTRACTORS

Some collection systems repairs and most new construction are contracted out to private firms. Managers of collection systems should establish good relationships and effective communication with private contractors. It is important to contact on-the-job consultants, contractors, and supervisors regularly and exchange business cards and telephone numbers with all parties involved. Familiarity with the scope of the construction is essential.

The contractor may be aware of the nature of a particular project but not know its effects on the remainder of the wastewater collection system. Regular meetings with the contractor can be held to communicate any concerns, needs, or reservations regarding the project.

Access should be provided to all public information, records, plans that are relevant to the project, and volunteer information that could be helpful to the contractor. It is important to monitor contractor activities and keep home telephone numbers available in the event of an emergency.

After the contractor has completed contractual obligations to the agency, the project becomes the agency's responsibility. Warranty information and operation and maintenance manuals for new facilities are obtained from the contractor.

MEETING THE NEEDS OF THE MEDIA

Some agencies designate a single media relations person, while other agencies may be without such a policy and may have no objection to managers and others communicating with the media.

Managers must recognize the role of the media in reporting newsworthy events when the agency's activities affect the environment. This includes learning how to deal with reporters and how to become a valuable resource to the media. Understanding the media makes the agency's work easier. The following steps promote positive media relations:

- Get to know media representatives before emergencies occur.
- Show respect for their deadlines.
- Provide reporters with appropriate background information.

- Be available to reporters.
- Be honest.
- Know how to write effective press releases and conduct effective press conferences.
- Read published news stories and listen to reports broadcast on television or radio about the facility (and ensure the information is correct).

PUBLIC RECORDS

Many public inquiries are connected with bills and are handled by the customer service department. However, homeowners, real estate agents, lawyers, contractors, plumbers, and others frequently contact the collection systems administration for information on sewers. The public typically seeks information on the following items: location, depth, and size of sewers; location of laterals that service private property; and availability and ability of sewers to service proposed residential, commercial, or industrial projects. The location and description of sewer easements on private property and other general information is also sought by the public. Up-to-date, legible, and accurate facility plans and records should be maintained. In addition to making plans and records available to the public, collection systems personnel should be able to explain and clarify the information. Be willing to discuss the availability and location of records including how, when, and where these records can be viewed.

JOB PROCEDURES

The collection system agency's relations with the community can be enhanced by informing the public of activities that will affect the community. If a sewer project involves excavating a street, rerouting traffic, and disrupting the normal activities of an entire neighborhood, a public meeting helps clarify the effects of the work. The contractor can prepare handouts describing the project and its timetable, and the collection systems manager and other involved agency personnel can explain the nature of the disruption and answer any questions.

If activities are confined to a small section of the roadway, direct personal contact with the affected property owners is important. Going door to door to explain the nature of the proposed activity helps gain acceptance of the project.

On-time job completion reduces disruption in neighborhoods and allows prompt return of normal activity. Proper and efficient job execution influences early job completion.

The collection systems manager and contractor should make every effort to minimize the disruptive nature of the work-related activities discussed in the following sections.

TRAFFIC. Traffic disruption can be kept to a minimum by prominently displaying warning signs near roadway interference. Local traffic regulation requirements may mandate the presence of a member of the local police department's traffic division.

NOISE. Noise from the project during normal working hours typically is tolerated by the community. Unless the job requires a 24-hour effort, activities involving motorized equipment should be confined to daylight and early evening hours (typically 8 a.m. to 8 p.m.). Take time to establish an adequate traffic control plan to accommodate the public and emergency vehicles.

ODORS. The agency should investigate any complaints from the neighbors concerning objectionable odors that are detected during construction or repair. Although people tolerate inconveniences and disruption in their daily activities if they understand the need for the project, most will not tolerate objectionable odors. Appropriate steps can be taken to resolve odor problems.

SAFETY HAZARDS. Any construction project in a community represents a physical change in the neighborhood and may necessitate vehicle traffic rerouting. Equipment on the job site often attracts youngsters. Changes such as these represent a potential public safety hazard. It is important that collection systems crews make every effort to warn the public of potential hazards and reduce safety problems. Supervisors and crews can anticipate occasional onlookers and possible illegal activities by some members of the community and make every effort to guard against these activities.

WRITTEN INFORMATION

Providing literature, pamphlets, newsletters, or information sheets to the public to help explain agency activities and responsibilities promotes positive government and community relations. This information may be included as an insert with the agency's bill or pamphlets and other literature distributed at community events, libraries, schools, or shopping malls.

Contact with the public before the start of a project is essential. Communities can be provided with written notices of impending projects that will affect them. These notices might include a project description, location, timetable, indication of the likely disruption, contractor's name, and agency staff contacts and telephone numbers. Personal communication with people in the neighborhood can be combined with the written notice. Personal contact with business and property owners and community leaders demonstrates the agency's sincerity in establishing positive public relations.

An agency "hotline" telephone number should be established to field inquiries or complaints, with the goal of addressing the complaints. This hotline number would also be displayed on any job site. All pumping stations and other agency facilities should certainly have an identification sign with an emergency phone number.

Chapter 6
Construction Contracting

This chapter provides an overview of public works construction contracting for wastewater collection systems and also provides sources of more detailed information. Significant information and guidance are also available from the following organizations: Construction Specifications Institute (CSI), Engineers Joint Contract Documents Committee (EJCDC), the American Public Works Associations (APWA), the American Institute of Architects (AIA), the Associated General Contractors of America (AGC), the National Utility Contractors Association, and the American Society of Civil Engineers (ASCE).

CONSTRUCTION PROJECTS AND PROJECT MANAGEMENT

Wastewater collection systems construction projects involve complex processes. Even small projects involve different skills, materials, equipment, and numerous operations. The construction process is subject to variable and unpredictable factors. For example, the aspects associated with different construction sites, such as subsoil conditions, surface topography, weather conditions, transportation systems, material and equipment supply, availability of skilled labor, local subcontractors, and many others contribute to the complexity of construction projects. Therefore, to some degree, every construction project is unique.

A construction project typically involves many parties, including the owner, design engineer, prime contractor, subcontractors, manufacturers, material and equipment suppliers, inspection and testing laboratories, and surety and insurance companies. The owner of the structure finances the project. The structure is designed in accordance with applicable codes and standards and follows state and administrative directives regarding advertising, bidding, contracting, and administration of the construction process. For competitive bidding, the owner defines the scope of the project, describes the materials and tolerances, develops complete and detailed descriptions of the proposed work, and describes the conditions by which the work will be performed. These elements are described in drawings and specifications prepared by the design engineer and are referred to as construction contract documents.

Construction drawings illustrate the extent and arrangement of the components of the structure. The specifications describe the materials and workmanship required. Construction contract documents are either prepared in house or subcontracted to an independent design engineer, depending on the complexity of the project and the owner's in-house design capabilities.

Project management is the overall control of design and construction to achieve an optimum balance between quality, schedule, and cost. Project management does not exert a constant, uniform influence over these attributes throughout all phases of project delivery, however. Project management influence is greatest during definition and preliminary planning phases and tends to decrease rapidly as the process continues toward design and construction. Major decisions concerning overall project size, scope, complexity, location, and performance characteristics are made during the early stages of the project. After the early stages of the project, firm commitments of project resources have to be made. As the owner commits resources to the project (such as acquiring facility sites, engaging consulting engineers for detailed design, and hiring contractors to perform construction), it becomes more difficult and expensive to make large changes to the project plan. As a result, proper planning and decision making are extremely important during project definition and preliminary planning.

Significant software is available to provide the tools and resources for this project management function. These project control systems include scheduling, estimating, and providing information management to control the budget, schedule, and quality of the project.

COMPETITIVE BIDDING

Laws or regulations typically mandate the competitive bidding of utility projects, ensuring value to the taxpayers and fairness in the expenditure of public money. Public administration of bids typically is required for publicly funded work. Project bids, which are based on construction contract documents, are designed to receive cost-competitive bids and thereby provide the owner with a constructed project of specified quality at the lowest possible price.

Special laws such as building codes, environmental regulations, building permits, safety and health regulations, zoning regulations, and licensing for contractors protect public health and safety and prevent fraudulent acts by owners and contractors. Contractors should be informed of these requirements in the bidding process.

FORCE ACCOUNT CONSTRUCTION

Construction projects involving repairs, small jobs, and emergency cases are often administered through a method called force account construction. Under this system, the owner acts as the prime contractor responsible for coordinating the project and either subcontracts work to specialty contractors or elects to do the construction with existing work crews.

CONSTRUCTION CONTRACT DOCUMENTS

The construction process typically includes documents that are essential to the bidding. These documents include

- Invitation to bid,
- Instruction to bidders,
- Bid form (proposal),
- Agreement,
- Bond forms,
- Insurance,
- General conditions,

- Special or supplementary conditions,
- Technical specifications,
- Drawings, and
- Addenda.

Other items detailed in the construction contract or contract clauses include methods and procedures for awarding the contract, notice to proceed, progress payments, retainage, acceptance and final payment, contract time, failure to complete work on time (liquidated damages), extension of time, warranty period, changes and change conditions, suspension of work, change orders, claims and disputes, indemnification, rights and responsibilities of the owner and contractor, duties and authorities of the engineer, arbitration, termination of the contract, subcontract provisions, and purchase orders. Standard contract forms and agreements are available through various professional organizations, such as EJCDC, AIA, ASCE, and AGC.

When a conflict arises in a contract, specific provisions have control over general provisions (that is, typed provisions override printed provisions). Ambiguous terms are interpreted in favor of the contractor, who had no role in drafting the document. If a conflict arises between drawings and specifications, specifications typically take precedence.

Claims for additional remuneration in construction arise when changes in conditions require additional contract time or cost. Most claims result from changes, delays, errors, or omissions in design; suspension of work; abnormal weather; or changes in subsurface conditions.

VALUE ENGINEERING

Value engineering typically involves engaging a separate engineering group to examine a project plan (or conceptual design) to question whether it will in fact deliver the value that the owner wants in a cost-effective manner, and also to consider whether other alternatives will prove more reliable and cost effective. The process forces the owner to question and determine the real need. It also assists in identifying the main advantages and disadvantages of alternative engineering options. Properly conducted value engineering studies should end with general agreement on the best engineering design for a particular owner, based on whatever the owner values most.

Value engineering can be introduced during the construction phase of a project. Because it is more difficult and expensive to change plans after significant resources have been committed to an existing plan, value engineering has been found to deliver the most value for the least cost during the earliest stages of a project.

Many construction contracts with public agencies include value engineering incentive clauses. Value engineering uses the contractor's special knowledge to reduce the cost of a project without sacrificing quality or reliability. A clause of this type provides the contractor with an opportunity to suggest changes in the plans or specifications (such as changing, for example, from open-channel trenching to trenchless technology) and share the resulting savings.

TYPES OF CONSTRUCTION CONTRACTS

Although there are many different types of construction contracts, they may all be grouped within the following major divisions.

- Contracts for which the contractor is selected on the basis of lowest responsive and responsible competitive bidder. This procedure is typical for public works contracts. Typically, these contracts will be lump-sum and unit-price contracts.
- Contracts resulting from direct owner–contractor negotiations. These can be determined on any mutually agreeable basis, such as a lump-sum, unit-price, or cost-plus fee.

LUMP-SUM CONTRACTS. Lump-sum contracts stipulate a fixed price for project construction. The lump sum reflects the total cost of the contract project. The owner benefits most from this type of contract because the total cost is known in advance. However, lump-sum contracts limit the scope of work to be done because all quantities and complete plans and specifications must be accurately known when the bid is made. The contractor assumes full risk because the job must be completed even though the total cost of the work may be greater than the contract price. Lump-sum contracts are often used in wastewater collection system construction for pumping stations and other well-defined structures.

UNIT-PRICE CONTRACTS. Unit-price contracts offer advantages for projects involving quantities of work that cannot be accurately forecast. In this type of contract, the estimated quantities are compiled by the design engineer, and unit costs for carrying out the proposed work according to the contract documents are bid by the contractor. The contractor is obligated to complete the work at the quoted unit price per listed item. Because payments to contractors are based on units of work actually performed and measured in the field, the exact cost of the construction is not known until the project is complete. As a result, unit-price contracts require complete plans and specifications. The contractor assumes less risk with unit-price contracts than with lump-sum contracts. Unit-price contracts are often used in wastewater collection system construction for pipelines and appurtenances.

COST-PLUS CONTRACTS. Cost-plus contracts are used when the owner wants a particular contractor to do the work, when the scope and construction techniques are not known, or when expedited project completion is necessary. In a cost-plus contract, the owner pays all the costs plus an agreed management fee; thus, the contractor takes the least risk. The owner and contractor typically negotiate the contract scope and, based on preliminary drawings and specifications, the contractor estimates the price. This type of contract often is used in force account work such as emergency repairs for wastewater collections systems.

Examples of items typically included in the management fee are contractor's head office costs, rents, taxes, insurance, interest on financing for the project, head office supervision and control costs, and contractor's profit. Under this type of contract, the contractor has the least incentive to keep costs down, though a number of variations to this type of contract have been developed to give the owner some cost control and provide incentives to the contractor to reduce costs. These variations include cost plus a percentage of cost, cost plus a fixed fee, cost plus a sliding fee, and guaranteed maximum cost plus cost savings sharing incentives.

BONDS

A bond is a legal document that binds a third party to the contract to provide assurance that the contractor will perform the agreed service. Under the bond, a party known as the surety assumes liability for the failure and debt of another. Bonds typically required in construction include bid bonds and contract bonds (performance bonds and payment bonds). Bid bonds assure the owner that the bidder, if selected for award, will sign the contract. Contract bonds guarantee that the contractor will complete the work according to contract documents. If the contractor fails to fulfill the contractual obligations, the surety completes the contract and pays all the costs up to the face value of the bond. Standard bond forms are available from many different organizations, such as CSI and EJCDC.

Construction contracts provide that the prime contractor warrants the work against faulty workmanship and materials for some period after project completion. One year is a typical warranty period, although periods of up to 5 years or more are typical on certain projects such as utility construction. During the warranty period, the contractor remedies construction defects at no cost to the owner. To ensure compliance with the warranty, a type of surety bond called a maintenance bond may be required. A maintenance bond requires the surety, if necessary, to correct defined defects in the contracted work that occur within the warranty period.

CONSTRUCTION INSURANCE

Utilities and states typically have specific insurance requirements in their regulations to protect the public; these are incorporated into the construction contract. For example, construction contracts typically require the contractor to provide coverage for workmen's compensation insurance, the contractor's public liability and property damage insurance, and the contractor's contingent liability insurance. Some contracts require the contractor to provide property insurance to protect the project and liability insurance to protect the owner. Many contracts require the contractor to protect the owner and design engineer by accepting any liability that may be incurred because of operations performed under the contract. Insurance certificates must often be submitted as proof of insurance.

QUALITY ASSURANCE, QUALITY CONTROL, AND WARRANTY ADMINISTRATION

Quality assurance (QA) is the sum total of management actions, strategies, and reporting functions required to assure the project owner that a constructed facility satisfies contract requirements and appropriate technical standards. Quality control (QC) is the inspection and testing function required to determine objectively the standard to which work is being performed and facilitate the reporting of results to management. While QA is a management tool, QC is a production tool. Many owners will hire an engineer to develop and manage a QA/QC program. Some agencies have their own QA/QC programs and hire inspection and testing agencies to carry them out. Others require the contractor to develop a QC program, and then assign a person or a third party to provide QA oversight. Elements of the QC program include project meetings, review and approval of submittals, field inspections, materials sampling and testing, and review and approval of record drawings. Systematic documentation and filing are the most important parts of a QA/QC program.

As construction proceeds, testing of individual components is performed, according to manufacturer's recommendations and recognized industry practices, to ensure that no damage will occur during larger scale testing of systems made up of assembled components. After construction is complete, total system testing, start-up, and commissioning begins. System testing proceeds as required by the manufacturer's recommendations, engineering specifications, and recognized industry practices. Results of these tests are documented carefully and systematically. Adjustments and corrections that may require engineering approval are performed as needed and carefully documented.

After all systems have been tested independently, the complete facility is ready for start-up. Start-up is performed in accordance with an operations and maintenance (O&M) manual, which incorporates all relevant data from equipment and instrumentation manufacturers, plus information provided by the engineer describing how the entire facility is designed to be operated and maintained. After start-up, the facility is tested and calibrated for optimum performance and efficiency. In some cases, the engineer may order modifications to the facility to satisfy performance requirements or improve efficiency.

Following commissioning, a final inspection is made, and job files, shop drawings, and O&M manuals (updated to include results of all facility calibration tests and any changes made during commissioning) are turned over to the owner.

A warranty is a guarantee that a facility will function according to stipulated and agreed performance criteria for a stated period of time. Warranties also provide for a remedy when this does not occur. Typically, they require the contractor to return to a facility and fix or replace any item that is not functioning to the owner's satisfaction. The contractor typically backs up the warranty with warranties from equipment suppliers. When a potential warranty issue has been identified, it is essential that facts be documented carefully to avoid acting on incorrect assumptions about who is responsible for a given situation.

SAFETY CONTROL METHODS AND PROCEDURES

The Occupational Safety and Health Act established a nationwide safety program for workplace safety and health. Employers are required to promote a work environment free from recognized health and safety hazards and to meet Occupational Safety and Health Act requirements. Construction contracts contain articles that require the contractor to conform to all applicable laws, ordinances, rules, and regulations related to safety. The contractor must also comply with any safety program policies instituted by public and local governments.

ONE-CALL SYSTEMS

Underground utilities pose significant problems during construction because information about their location may be lacking. Many communities have set up what is referred to as a one-call or dig-safe system so that one phone call locates all recorded utilities in the area of construction and notifies appropriate utility companies. The APWA's *One-Call Systems Manual* (1978) and *How to Form Utility Location and Coordination Committees* (1977) can be used to locate utilities. Uniform color-coded cards are available to standardize marking locations, changes in direction, and dead ends of buried lines. Proper utility location helps avoid accidental damage and service interruptions.

Subsurface utility engineering is the process of locating and depicting the location of underground utilities according to an industry-established and -defined level of quality. This practice has been promoted by the Federal Highway Administration as a cost-effective means of providing accurate underground utility information for design, which otherwise may cause costly construction delays.

ALTERNATIVE CONTRACTING METHODS

There is a growing trend to consider alternative contracting methods to build public infrastructure projects. The traditional project delivery approach is design, bid, and build. However, there are other methods that may be less costly, less prone to litigation, and completed in less time. These contracting methods include construction management and design build.

The benefits to the owner follow:

- Project delivery time reduced—the potential exists to reduce the total project delivery time from inception to completion. This is accomplished by phasing design and construction, reduction in schedule delays, and collaborative decision making by the design and construction personnel.
- Quality improved—the integration of design and construction personnel provides many opportunities to improve the quality of the constructed facility.
- Project cost reduced—the potential exists to reduce the project cost. This is possible through the reduction of project delivery time, the reduction in construction cost, and the reduction of claims. Influence on the cost of a project is most dramatically affected at the beginning of the project. Bringing together the design and construction personnel at the beginning provides the largest potential for construction cost savings.

CONSTRUCTION MANAGEMENT. The construction management approach to contracting typically involves the construction manager assuming the role of the general contractor and includes the provision of design-phase services. Frequently, this approach is used to "fast track" a project by phased construction. The owner, engineer, and construction manager work together to control and balance the cost, quality, and time from inception to completion of the project. The purpose of adding the construction manager to the team is to deliver the maximum value of a project within the most economic time. Construction management firms should be selected based on qualifications, approach, and price. The process involves prequalification and shortlisting, preparation of requests for proposals, submission of proposals by prospective firms, team evaluation based on a defined rating system, interviews, follow-up evaluation of the candidates, decision notification, contract negotiation and approval, and notice to proceed.

DESIGN BUILD. Design build replaces the traditional method of awarding separate contracts for design and construction. Design build is a method of project delivery in which a single entity provides all of the services necessary to both design and construct the project. This single point of responsibility fundamentally distinguishes the design-build format from other forms of project delivery. A single source of responsibility reduces the owner's administration and management burden, protects the owner from liability, and reduces disputes concerning responsibility. Planning and procurement for successful design-build delivery differs in several ways from traditional project delivery. Strong owner involvement is required at the planning stage to determine the performance characteristics of the planned facility with sufficient clarity to establish fair competition. The procurement process must also allow the designers sufficient flexibility to consider economical technical alternatives, and must require designers to guarantee the performance of their facilities. It must also allow a basis for fair comparison between alternatives and provide a remedy when performance criteria are not met. Because teamwork is an important part of the process, owners must take great care to select qualified, functional design-build teams. The selection process described for construction management should also be used for the design-build format.

PARTNERING

Owners have become frustrated with disputes and the resultant effects on cost and schedule. Partnering has evolved as a concept to help head off disputes before they become an issue. Partnering is a method to open lines of communication and improve the process.

The Construction Industry Institute has defined partnering as "a long term commitment between two or more organizations for the purpose of achieving specific business objectives by maximizing the effectiveness of each participant's resources. This requires changing traditional relationships to a shared culture without regard to organizational boundaries. The relationship is based on trust, dedication to common goals, and an understanding of each other's individual expectations and values. The expected benefits include improved efficiency and cost effectiveness, increased opportunity for innovation, and the continuous improvement of quality products and services."

The partnering process involves a partnering development meeting, a joint partnering workshop, periodic evaluations, required escalation of an issue, and final evaluation and project commemoration.

REFERENCES

American Public Works Association (1978) *One-Call Systems Manual.* Chicago, Ill.

American Public Works Association (1977) *How to Form Utility Location and Coordination Committees.* Chicago, Ill.

SUGGESTED READINGS

American Society of Civil Engineering (1991) *Avoiding and Resolving Disputes During Construction.* New York, N.Y.

Barrie, D.S., *et al.* (1978) *Professional Construction Management.* McGraw-Hill, Inc., New York, N.Y.

Clough, R.H. (1981) *Construction Contracting.* John Wiley & Sons, New York, N.Y.

Clyde, J.E. (1983) *Construction Inspection—A Field Guide to Practice.* John Wiley & Sons, New York, N.Y.

Construction Specifications Institute (1990) *Manual of Practice.* Alexandria, Va.

Del Re, R., and McKittrick, H.V. (1985) *The Rule of the Resident Engineer.* Am. Soc. Civ. Eng., New York, N.Y.

Lambert, J.D., and White, L. (1982) *Handbook of Modern Construction Law.* Prentice-Hall, Inc., Englewood Cliffs, N.J.

Meier, H.W. (1989) *Construction Specification Handbook.* 4th Ed., Prentice-Hall, Inc., Englewood Cliffs, N.J.

Chapter 7
Training

Consumers today expect greater service from both the public and private sectors. As a result, increased service and responsiveness will be demanded of collection systems personnel, who often find themselves making repairs or planning decisions that cost thousands of dollars. Thorough basic training helps ensure the efficient use of system personnel and facilitates a safe working environment. In doing so, it also improves service competitiveness, reduces turnover, and lifts morale.

Learning is the process of changing behavior, and training provides experiences to aid in learning. An organized training program is a necessity, regardless of agency size. Managers have not only a responsibility under the law to provide certain training: they have a broader, ethical responsibility to do more than the legal minimum.

Typically, 3 to 5% of the gross budget is set aside for training expenditures, which include employee wages while attending classes, seminar fees, trainers' salaries, and training supplies. Not all classes should be held during working hours: training is often more effective when employees invest some of their own time. Top management should support training and include preventive maintenance as a priority.

IMMEDIATE TRAINING OPPORTUNITIES

ORIENTATION TRAINING. It is a good idea to conduct an employee orientation in the first week of employment or reassignment. This orientation can include how to fill out a time sheet, when and where to take breaks, and how to operate a jetting machine. An immediate employee indoctrination program helps increase morale. In addition, the safety emphasis catches employees' attention, and fewer bad safety habits may develop as a result.

TRAINING BENEFITS. Collection systems employees are exposed to numerous challenging conditions, and adequate training enables employees to meet these challenges. Managers will be able to identify immediate quantifiable benefits from training. Examples of such tangible benefits are a reduced number of serious injuries, lowered legal fees, lowered insurance rates, the fulfillment of statutory legal requirements such as Occupational Safety and Health Administration right-to-know legislation, decreased sick leave hours, reduced disability claims, fewer equipment accidents, greater opportunities for employees (promotion and increased salary), lower electrical and equipment costs, and compliments from citizens.

TRAINING TOPICS. Some topics for which related training programs are likely to provide immediate (and long-term) positive results include

- Answering the phone;
- Handling a complaint call;
- Entering private property;
- Implementing effective public relations;
- Implementing emergency response protocol;
- Identifying a confined space;
- Entering a confined space;
- Using a confined space entry permit;
- Locating traffic safety aids;
- Using traffic control permits;
- Learning about common chemicals and their proper use;
- Wearing proper attire for a job and climate;
- Using safe electrical practices;
- Performing proper stop, lockout, and tagout of equipment;
- Using compressed fluid or gas equipment properly;
- Learning cardiopulmonary resuscitation;
- Learning first aid;
- Responding to workplace accidents;
- Using motor vehicles properly;
- Using protective clothing and other typical safety gear;
- Keeping emergency maintenance records;
- Operating a basic pumping station;
- Learning about types and uses of lubricants;
- Learning basic hydraulics (siphons/overflows, pipe flows, and velocities);
- Learning the importance of collection systems;

- Reading construction plans; and
- Inspecting sewer connections.

LONG-TERM TRAINING OPPORTUNITIES

The long-term opportunities of a training program are numerous. To understand these benefits, it is necessary to look at the types of behaviors employees may be required to change.

- Cognitive—actions that involve knowing, understanding, or applying information; remembering or using facts and ideas.
- Affective—expressions of feeling or interest; the adoption of an attitude or belief; the motivation to do something.
- Psychomotor—physical action, speed, or agility; the performance of a particular body movement.

The application of techniques ("doing physical work") is often emphasized, but time is rarely spent teaching the basics of why and how processes work. Many employees learn through repetition. For example, when an employee is instructed to turn on a switch when a wet well fills to a certain depth, he or she is rarely told why. The employee may have performed this task for years without understanding the reasons behind it. An employee with excellent applications skills but who lacks a basic understanding of the process may have limited troubleshooting skills. In the preceding example, what would happen if the switch failed to turn the pump on? If the employee had performed the function for 3 years and the pump had never failed, it is possible that he or she would not know what to do. A balanced training program that includes foundation skills positions the workforce for safe, efficient, courteous service to the community or industry.

The following are training opportunities over the longer term:

- Using proper excavation techniques in varying weather and soil conditions,
- Planning crews,
- Selecting pipe material and technique,
- Handling problem employees,
- Conducting a performance review,
- Measuring performance,
- Establishing performance expectations with measurement criteria,
- Deciding when to replace rather than repair equipment,
- Spotting typical design deficiencies in collection systems (operation and maintenance considerations during design review),
- Learning how to measure wire-to-water efficiency,
- Using odor control methods,
- Learning how to make an impressive public presentation,
- Learning how to deal with the media during a crisis,

- Learning corrosion prevention,
- Learning about the effects of hydrogen sulfide and accident prevention,
- Using proper gas-sampling techniques,
- Learning methods of pipe repair,
- Understanding the effect of seasons on the maintenance cycle,
- Selecting specific equipment or procedure skills,
- Sizing and selecting a replacement pump efficiently,
- Learning the benefits of records and reports,
- Setting up a blanket purchase contract,
- Using a complaint incident report,
- Learning the differences between time and materials versus lump-sum contracting,
- Learning advanced instrument repair,
- Using personal computers including spreadsheet and database programs,
- Practicing proper crew routing,
- Using a comprehensive maintenance program properly,
- Setting up a contract for replacement or repair, and
- Administering a contract.

TRAINING PROGRAMS

Total management commitment is a key to a successful training program. Programs should be packaged so that top managers are behind the task. Resources needed to set up a training program include

- Funds for training materials, training aids, audiovisual equipment, trainers, labor, and training space;
- A training coordinator; and
- A training committee (the dedication and commitment of several employees representing a cross section of the workforce is needed; these individuals set goals, evaluate milestones, and conduct other organizational activities).

The following steps will help start the training program:

- Formulate four to six clear, concise program objectives and goals.
- Establish a training unit. In a large agency, this may include several full-time people who report to the department manager. This group serves the various collection systems groups and focuses on goals.
- Develop a detailed action plan that centers on achieving the objectives and goals. Include short- and long-range resources for accomplishing this.

- Establish a formal feedback and participation mechanism to foster ownership by everyone.
- Conduct internal public relations efforts and communicate with the department. For example, distribute internal news releases and newsletters and hold lunchtime discussions and toolbox talk. Be innovative and use imagination. Pay attention to people and they will respond favorably.

TRAINING COMMITTEE. A training committee chairperson from a non-management position fosters employee acceptance. The training committee develops goals and objectives for the overall training program. These goals require specific measurable criteria and are designed to enhance skills development and improve compliance performance, job performance (cost reduction), employee morale, public relations, and education for staff.

After these preliminary goals are established, they are presented to the employees through the internal public relations program. After receiving feedback, final goals are established.

The training committee is then responsible for establishing its own purpose, policy, and goals. A specific policy statement is approved by management. The training committee meets briefly at least once each month and holds a quarterly meeting focusing on strategic objectives and long-term planning.

TRAINING TECHNIQUES. The learning process works better when training aids are used to visualize concepts and reinforce verbal instructions. Whether learning in a classroom, in the workplace, or in the open, the use of visual (sensorial) training aids greatly improves the trainee's ability to retain information.

Typical training aids include equipment demonstrations, hands-on experience, slide and overhead projector presentations, handouts, and programmed notes. Which training aids are used depends on the audience and complexity of the message. Objectives, audience size, location, budget, preparation time, amortization (reusability), and complexity of message are all factors to consider when preparing training aids.

Classroom training combined with guided hands-on experience is most effective with collection systems personnel. For example, the proper use of a self-contained breathing apparatus is retained best if employees properly prepare for, put on, use, and restore or recharge the equipment. Periodic refresher exercises should also be performed.

CASE HISTORY EXAMPLES. A number of municipalities have moved forward with very innovative training approaches. For example, we have known for years that "in the field" training experiences, under near real conditions, are the best experience. Retention is greatly improved. The following

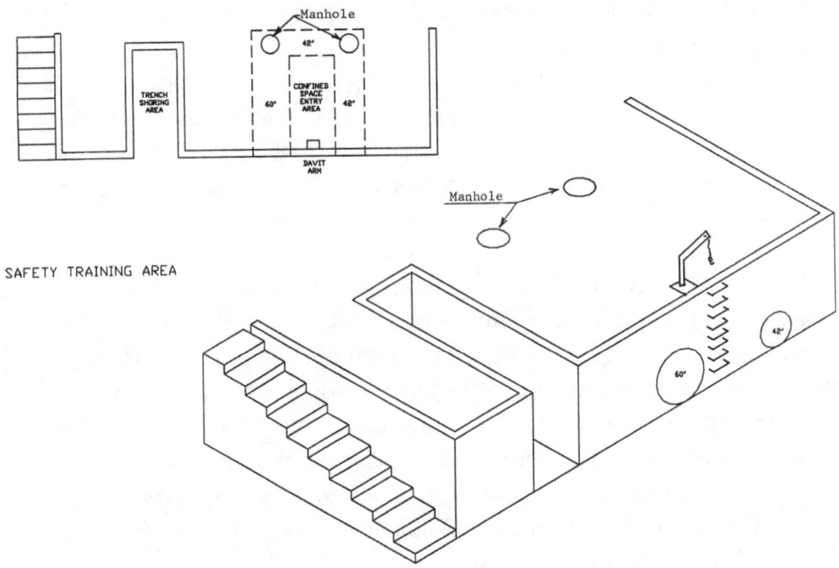

Figure 7.1 Safety training area.

two training facility examples reflect state-of-the-art collection system training and serve as examples of how training can be approached in a more meaningful manner.

A training facility built by the City of Columbus, Ohio, Department of Public Utilities (see Figures 7.1 and 7.2), located on the grounds of its Sewer Operations Center, provides a simulated environment in which employees can learn proper confined space entry. The arrangement has helped the divi-

Figure 7.2 The city of Columbus, Ohio, Sewer Maintenance Operations Center Training Facility for confined spaces.

Figure 7.3 Frontal view showing trenching.

sion comply with applicable safety regulations. Facility training stresses all the usual confined space scenarios but also provides experience with trenching and shoring. The concrete trench shown in Figures 7.3 and 7.4 provides a useful means to train employees without the possibility of a cave-in. All methods, whether hydraulic or timber, can be demonstrated and exercised.

The facility's controlled environment allows training in entering and exiting confined spaces—with openings as small as 0.5 m (22 in.)—without the

Figure 7.4 Frontal view showing the trench shoring area from lower level.

danger of falling. The simulated condition lessens the risks from dangers found in actual collection systems, such as toxic gases. The location is in the central yard, close to all equipment and facilities. Trainees are not in danger of traffic, and can learn from their mistakes. In addition, the facility provides supervisors with the flexibility to schedule training as needed. Finally, it allows the division to test new equipment before using it in the field.

The layout of a smaller confined space training facility, built by the Metropolitan Council Wastewater Services in St. Paul, Minnesota, is shown in Figure 7.5. The unit is built into a hill, where the "cattle" entrance is exposed on the side of the hill. An instructor can enter from this entrance and open a steel door. A light and power for camera equipment are available. The trainee enters from above, using techniques first taught in a classroom.

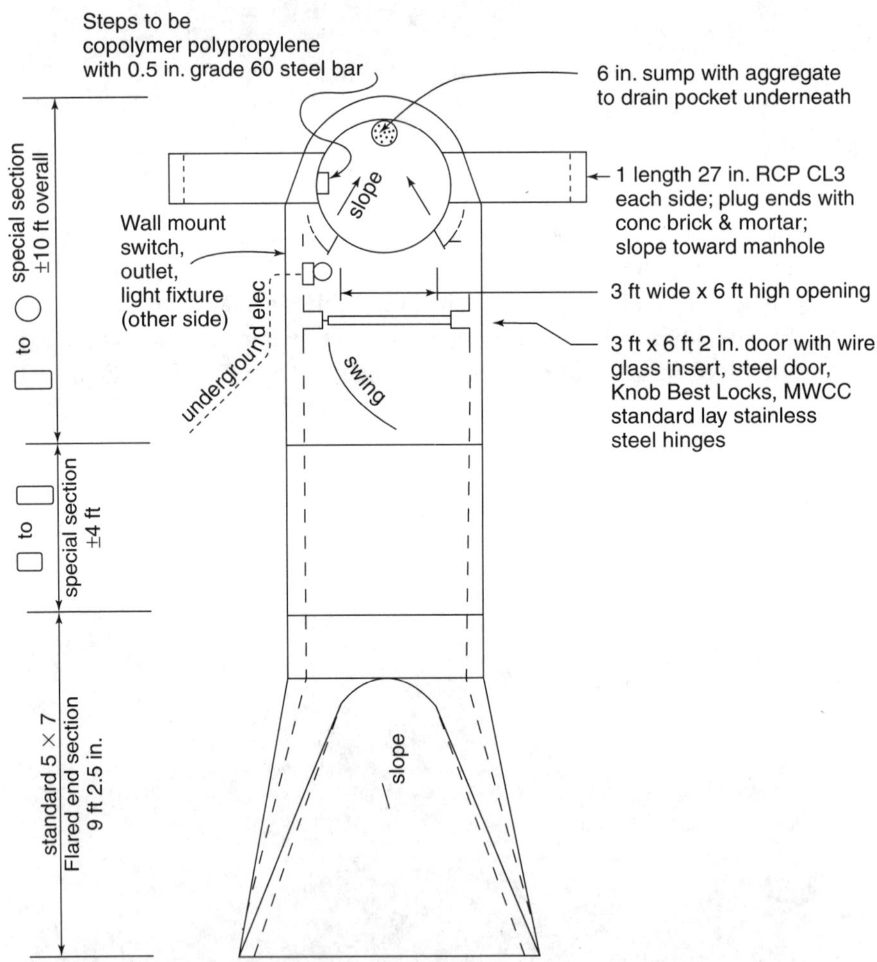

Figure 7.5 Sample process and instrumentation diagram (ft × 0.304 8 = m; in. × 25.4 = mm).

Regardless of training needs, these facilities are excellent examples of what agencies can do to improve the learning experience, educate more effective service providers, and reduce potential harm to trainees.

FOLLOW-UP. It is important to carefully track employee development and participation in training programs. Tracking should include periodic consultations with the employee during which goals are established and reviewed. Accomplishments are highlighted and goals adjusted based on individuals' own goals, their progress, and the employer's changing needs. Also, continuing education units (or CEUs) can be administered. One CEU is equal to 10 contact hours. Often, submitting an outline of the training program to the state environmental certification department will allow employers to issue recognition in the form of contact hours or CEUs, which can then be used by the employee for certification. Some states require that each person certified must obtain a certain number of contact hours of training each year to maintain certification.

A personal computer eases the tasks of tracking progress and inventorying resources. At a minimum, the database for each employee should include a social security number, title, the area of the employee's workstation, type of license held, and all courses completed.

The database for the training includes

- Program,
- Class,
- When the class was held,
- Who attended,
- Duration of the class,
- Author of the class or instructor,
- Test given,
- Results,
- Pages of text,
- Number of overheads,
- Number of slides,
- Whether student notes were given,
- Whether an instructor's guide was used, and
- Continuing education units awarded.

Components of a library database are titles, authors, subject matter, and the location of resources.

The database and other resources can be kept in an area at the central yard called the resource center, where audiovisual equipment, library resources, training records, and classrooms are maintained. Training materials should be available to personnel for reference during the typical conduct of their duties. Quarter section maps, plans, and specifications can also be housed in the resource center. Classrooms can serve as conference rooms when classes are not in session.

SUGGESTED READINGS

Horton, D. (1983). *Advanced Instructional Technology.* U.S. EPA, Columbus, Ohio.

Horton, D. (1983). *Basic Instructional Technology.* U.S. EPA, Columbus, Ohio.

National Training Operations Technology Center (1980) Water Quality Control Instructional Materials. Cincinnati, Ohio.

Chapter 8
Capital Asset Management

OVERVIEW

Increasingly, professionals and policy makers are becoming aware that infrastructure exists as an interconnected system and that efficient functioning of the system is crucial to environmental quality, economic well-being, and quality of life. The wastewater collection system, which includes sewers, pipelines, conduits, pumping stations, force mains, and other sewer-related facilities, is one of the largest infrastructural assets a community owns. Both public and private wastewater companies are being challenged to meet new state and federal regulations, increased operating costs, and the expectations of customers for higher standards, while facing a deteriorating infrastructure and a decreasing budget. Sewer systems degrade continuously, and a plan is necessary to effectively manage this degradation and prevent collection systems failures caused by pipe aging. Sewers that were built in the 1940s and designed with a 50-year useful life (WPCF, 1982) are now beginning to fail. In some systems, these older sewers represent more than half the system. Deferred maintenance will only increase costs for later generations and may cause catastrophic results. Figure 8.1 is an example comparing structural needs and investments for an aging collection system in a large midwestern city. It is easy to see how costs can become uncontrollable if infrastructural investments are not made. Preventive maintenance and rehabilitation extend the useful life of systems and lead to associated costs that are significantly less than costs for emergency repairs or replacement.

Currently, utilities are moving toward using a cost-of-service or activity-based cost (ABC) rate structure. By recognizing the annual revenue received from each customer and monitoring operating expenses of the infrastructure delivery system, a municipality can value its asset base and perform customer profitability, return-on-investment, and return-on-assets analyses, similar to analyses conducted by for-profit organizations.

Increasingly, utilities are turning to asset management planning to manage their collection systems. An asset management plan is a framework that brings all the key components of running a wastewater utility into a strategic business plan. In the capital costs portion of the plan, the starting point is the target standard of service to be provided the customer.

An asset management plan provides a comprehensive assessment of the

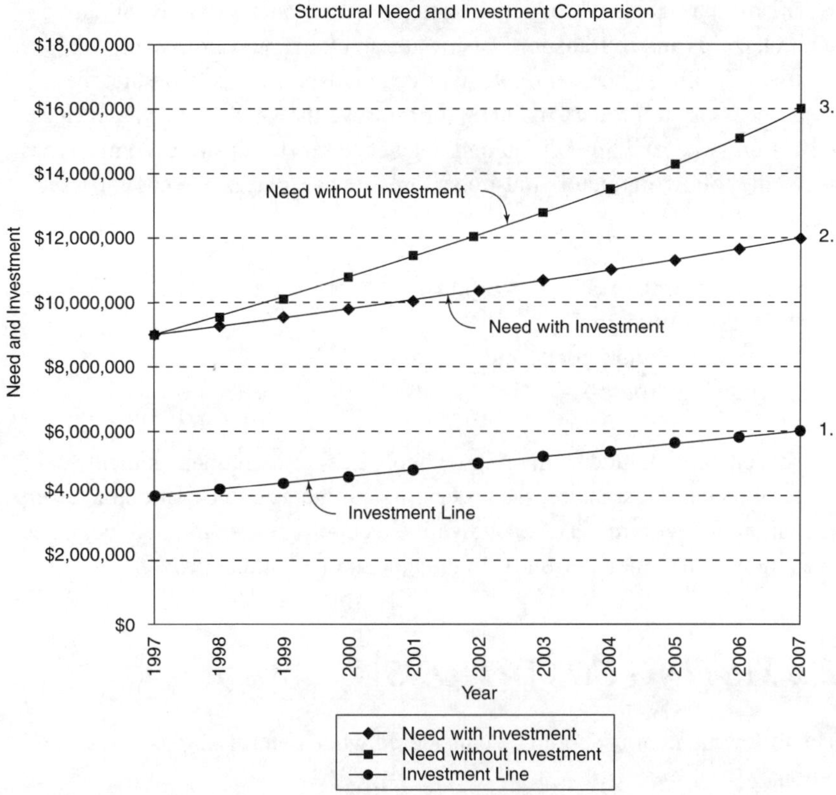

Figure 8.1 Example of structural needs and investments for a large midwestern city. Line 1 is the proposed structural capital improvement/repair investment of the collection system; line 2 is the structural need if line 1 investment is made; and line 3 is the structural need if no investment is made.

current condition of the utility and a projection of the cost required to meet future standards of service (Kingdon, 1995). Through financial modeling, an assessment can be made of rate changes needed to meet those standards. In particular, the asset management plan will allow the owner to establish trade-offs between rates and levels of service.

Capital asset management is based on the standard of service required by the customer. The capital asset management system provides a means to protect, maintain, or improve the asset value of a collection system with planned maintenance and repair based on predicted deterioration of the system. In either a private or public utility, key information is needed to manage costs through asset management planning (Kingdon, 1995), including current conditions and performance of assets; current operating costs; current financial position including revenues, balance sheet, and cash flow; required and anticipated future levels of service; and methods of measuring and monitoring performance of the system.

The primary goals of collection systems operation are to prevent public health hazards and to transport wastewater uninterrupted from source to treatment facility. The goal of capital asset management is to predict the costs necessary to protect, maintain, or improve the value of this asset (the collection system) while providing the level of service desired. To meet these goals, the following factors must be addressed as a part of the capital asset management plan:

- Determining existing conditions,
- Setting future goals,
- Attaining future goals, and
- Tracking progress.

This chapter includes emerging technologies for asset management. Because of its complex nature, the process for predicted deterioration of sewers is continually evolving. These innovative technologies should be recognized, even though they may currently be undemonstrated and unproven.

EXISTING CONDITIONS

The following factors should be considered when determining existing conditions:

- Maintenance, equipment, and organization;
- Data management strategy;
- Financial conditions;
- Activity-based costing;
- Useful life and depreciations;
- Life-cycle cost;
- Staffing;
- Existing collection system appurtenances;
- Customers;
- Funding and grants; and
- Fleet management.

MAINTENANCE, EQUIPMENT, AND ORGANIZATION. Chapter 3 addressed different variables that affect the size of the maintenance organization. It also provided information on inspection, regular maintenance, and emergency and repair crews. Historically, utilities have often taken a crisis management approach to maintaining their infrastructures. In this type of management system, problems are addressed as they occur. Little thought is placed on improving the way the maintenance department is managed and how maintenance work is performed. In many instances, problems may become extremely severe before they are detected.

Conversely, some utilities have taken a preventive maintenance approach. This type of program detects problems that already exist and prevents problems. The preventive maintenance approach includes programs such as sewer cleaning, televising manhole inspections, and planning of repairs. A preventive maintenance program is an effective way to keep the system in good repair and operation.

Using a predictive approach to sewer maintenance reduces costs by allowing the utility to predict expenditures in advance. This type of approach includes provision of financial resources to repair and replace sewers while protecting the asset of the collection system. Currently, the Water Environment Research Foundation (WERF, 1997) is beginning development of predictive tools for measuring the aging of wastewater collection systems.

Regardless of the approach used to define maintenance needs, each utility should establish system goals and performance indicators that can be used to evaluate the effectiveness of its management. It is becoming increasingly important to determine "how much maintenance is enough." Chapter 3 indicates there should be some reasonable balance between the cost of preventive maintenance and the derived benefit. The goal of capital asset management is to achieve this balance by maintaining a reasonable cost that is enough to protect, maintain, or improve the asset (collection system) value and facilitate performance of the collection system at the level of service desired by the customer.

DATA MANAGEMENT STRATEGY. The acquisition, analysis, and application of collection systems information is the foundation of an asset management system. A utility's ability to effectively manage its infrastructure is directly linked to its ability to maintain accurate and up-to-date information on the status and location of its facilities. At a minimum, this requires that information about changes made by field crews, such as operation and maintenance (O&M), be recorded and shared with staff maintaining the facility's maps, records, and engineering information. In a manual records environment, the best organizations may achieve this goal approximately 50% of the time.

A computerized maintenance management system (CMMS) is essential to proactive maintenance management. Today, most systems run on personal computers and have extensive graphical user interfaces for ease of use. Advances in computer technology have made the development and maintenance of databases cost effective for even the smallest utilities. Systems should be tailored to meet unique facility needs. A careful evaluation of specific needs, based on clear objectives, should guide the development of an effective system.

Specific categories of information necessary for computerized maintenance systems are facilities inventory, service work order, and maintenance history (Lior, 1996). Inventory data should be available within the CMMS so that jobs can be easily planned and crews can be sent to locations with the appropriate parts and materials. This information is important to the organization for other uses, such as asset management, capital planning, or hydraulic modeling. Historically, location and attribute data have been maintained on maps and drawings, inventory cards, or manual records. The data conversion activity will be the largest effort required of most utilities.

Information about a collection systems inventory can be linked to its geographic location through the use of identification numbers, geocodes (assignment of latitude and longitude coordinates to specify utility attribute data and street addresses), addresses, and global positioning system and state-plane coordinates. Computerized maintenance management system software should be evaluated for its ability to contain these geocodes, address data, and state-plane coordinates in facility records.

The primary function of most CMMS applications is service order processing. All service orders must be geographically referenced and should provide information about the facilities involved. Location plays a key role in scheduling and tracking maintenance service orders. Inspectors are dispatched to a street address to investigate service order requests, for example, or work is scheduled by geographic area. Maintenance history data are valuable for determining budgets, planning capital improvements, and developing preventive and predictive maintenance activities. These data include the facilities involved; the problem reported; activities performed; materials, parts, and labor required; and job site information. Geographic information systems (GISs) facilitate retrieval and use of this information by identifying specific facilities, locations, or attributes.

After maintenance history and facility inventory data are available through GIS, the data also can be reviewed in conjunction with other coverage found in the GIS. For example, a sewer's maintenance history can be overlaid with results from a hydraulic analysis to rank potential capital improvement projects and to protect assets.

Preventive maintenance activities for most facilities typically are planned by geographic area, maintenance history, or facility attributes. Retrieval of facility information by area within the city depends on geographically referencing facility inventories. For most organizations, maintenance history data are critical for evaluating the effectiveness of preventive maintenance programs.

Benchmarking is a tool that management can use for improving performance of collection systems (Allio *et al.*, 1994). Many large corporations have used benchmarking, and the fundamental management principles affecting productivity, profitability, or organization are as valid for wastewater collection systems as they are for industry. The benchmarking process is straightforward, as shown in the following six steps:

- Identify benchmarking functions;
- Identify performance criteria;
- Identify best-in-class utilities and measure best-in-class performance;

- Measure system performance;
- Close the gap between measured system and best-in-class performance; and
- Implement changes.

When identifying benchmarks, those functions that can be easily measured and that allow managers to exert an influence through changes in policies and procedures should be targeted. Identification of performance criteria should include consideration of both financial measures (such as return on assets) and strategic measures (such as customer-perceived quality). When current performance is measured, the organizations being compared should have the same performance variables. These performance indicators are sometimes needed for wastewater collection systems to justify budgets. Any improvement plans to close the gap between current performance and the performance of competitors must be weighed against the cost and investment required to achieve those results. Finally, implementation of any changes should consider the length of time needed to achieve results and include a system to monitor and reassess the results.

FINANCIAL CONDITIONS. Financial management is important for two significant reasons: a financial management process can provide local officials with a high level of confidence that resources are being used properly, and financial management reports can provide local officials with the information they need to evaluate utility performance and set a course for future action. Complete and timely financial reports are vital ingredients in successful management and policy making. These reports should cover such matters as local financial conditions, actual performance against budgeted expenditure and revenue targets, and the cost of providing specific types of service. Through sound accounting procedures and periodic audits of financial records and procedures, assurance is gained that the resources of the utility are properly safeguarded. The following section sets forth some basic procedures and practices and shows how the utility manager can match actual results of operations with planned or budgeted amounts.

Revenue Requirements. In order to adequately fund system development and O&M, and to maintain the ability to borrow, utilities must ensure that sewer rates generate sufficient revenue. Rates typically are based on the coverage requirements of bonds found in the utility's bond ordinance or bond resolution; that is, net revenue (defined as gross revenue less O&M expense, exclusive of depreciation) is 1.25 times the maximum (average) annual debt service of all outstanding bonds plus those to be sold. When net revenues fall below the coverage requirement, the utility is technically in default of its bond ordinance. Typically, a majority of sewer revenues are derived from water-utility-metered quantities and surcharges for wastewater strength, such as biochemical oxygen demand or suspended solids. Interest on investments, sewer taps, auction sales, and other reimbursements should also be considered.

Typically, bond issues for capital improvements are for 10 to 40 years, with most being 15 to 20 years with 10 to 30% bond coverage. These bond issues typically are revenue bonds (paid by customer usage). Other funding is also available from taxes. If revenue bonds are used, they should be properly structured so as not to count toward the local government's constitutional debt limit (for example, 2% in Indiana). It is wise to consult with experienced attorneys and rate consultants as early as possible when planning rate increases and bond issues.

Although user charges based on volume (assuming uniform wastewater characteristics) typically account for more than 90% of revenue requirements, other factors should be considered. A recent study (Woolpert Consultants, 1997) for a large sewer district in northern Kentucky, for example, identified several items that were not financially self-sustaining. Table 8.1 shows actual costs, existing costs, and proposed charges arising from the study. Many cities subsidize these types of charges with user volume charges.

Ad valorem property taxes are often used as the primary source for supporting a community's wastewater treatment facilities (WPCF, 1984). In some cases, *ad valorem* taxation can be combined with a service charge system to obtain the required operating revenues.

Table 8.1 Proposed permit and fee schedule.

Item	Present charge	Actual cost	Proposed charge
Inspection fee	\$0.55/ft[a]	\$0.75/ft	\$0.75/ft
Plan review fee	None	\$300	\$300
Sludge hauler fee	\$50/y	\$75/y	\$75/y
Sludge dumping fee	\$35/1 000 gal[b]	\$35/1 000 gal	\$35/1 000 gal
Pretreatment permit	\$650	\$1 200	\$1 200
BOD surcharge	\$0.069/lb[c]	\$0.201/lb	\$0.201/lb
Suspended solids	\$0.074/lb	\$0.12/lb	\$0.12/lb[d]
Total Kjeldahl nitrogen surcharge	\$0.059/lb	\$0.804/lb	\$0.804/lb
Grease trap fee	None	na	Violation fine
Capacity analysis	None	\$500	\$500
Violation fines	\$5 500	na	\$5 500
Telemetry charge	\$5 000	\$6 700	\$6 700
Minimum tap-in fee	\$500	\$1 800	\$1 800

[a] ft × 0.304 8 = m.
[b] gal × 3.785 = L.
[c] lb × 0.453 6 = Kg.
[d] Included with the BOD surcharge.

Projection of Revenue Requirements. Some utilities project revenue requirements only 1 year in advance, possibly because 1-year budgets typically are adopted. From a revenue-adequacy standpoint, projections beyond 10 years are of questionable value. When a utility adequately plans ahead, 5-year projections typically are sufficient to satisfy investors, bond-rating agencies, and others. This period presents a reasonable forecast of anticipated future revenue needs, thereby assisting management, policy makers, and the public in foreseeing potential problems.

Revenue requirements may be projected through the cash-needs approach and the utility approach, whichever is appropriate. In each case the O&M and debt-service costs should be identified and addressed separately according to state and federal regulations.

CASH-NEEDS APPROACH. The cash-needs approach (AWWA, 1990) projects revenue requirements according to the total cash needs of a particular projection period, and is generally used by government-owned utilities. Components of the cash-needs approach include O&M expenses, debt-service payments, contributions to specified reserves, and capital expenditures, but do not include depreciation losses.

O&M expenses generally include salaries, wages and benefits, purchased power and other services, rent, chemicals, other materials and supplies, small equipment, and general overhead. The debt-service component includes principal and interest payments on bonds, and may also include debt-service reserve requirements as established by the indenture. Other reserves are often required to provide for emergency repairs and replacements, as well as for routine replacements and extensions. Capital expenditures include routine replacement of existing facilities, normal annual extensions and improvements, and major capital replacements and improvements.

UTILITY APPROACH. Government-owned utilities under state regulatory jurisdictions are required to use the utility approach (AWWA, 1990) to project revenue requirements. In addition to the O&M expenses component the utility approach addresses depreciation costs, returns-on-rate base, and taxes or other municipal payments.

By including depreciation costs in revenue requirements a utility may have access to funds for improvements or replacements designed to protect the facility's systems from deterioration. New methodologies are being developed to predict future depreciation of collection systems. When these methods become available, asset management will become a more effective tool to maintain the useful life of the collection system. When calculating depreciation, the value of the system should include its full cost (including grants) or value. Also, if rates are tax based, a key element that local governments must keep in mind is the constitutional debt limit.

The return component of revenue requirements is based on a rate base, or the value of assets on which the utility is allowed to earn a return, which includes that portion of the utility's property which contributes to the utility's production or service. Revenues must be sufficient to cover interest payments on debt capital and provide a realistic rate of return.

Because regulations for determining total revenue requirements vary among the states, it is imperative that the utility investigate the policies of the regulatory body under whose jurisdiction it operates.

Test Year Period. Annual revenue requirements are determined using data from a test year (AWWA, 1990), a period of 12 months or more from which data are used in allocating costs and designing rates. The data may be historical, current, or projected. Utilities which operate inder the jurisdiction of state public utility commissions must follow particular state legislative and regulatory practices in setting test year periods. Table 8.2 (AWWA, 1991) shows a sample test year for cash-needs approach and utility approach comparison.

Revenue Basis. Many utilities set their rates according to records accounted on an accrual basis, with which a realistic assessment of profitability may be made. Many government-owned utilities, however, are obliged by bond indenture to maintain accounting records on a cash-receipts basis (or modified accrual basis), and must provide for the lag in cash revenues behind accrued revenues.

Table 8.2 Sample test year. (Reprinted from *Water Rates—M1*, by permission. Copyright © 1991, American Water Works Association.)

	Government-owned utility		Investor owned utility
Line number	**Cash-needs approach**	**Utility approach**	**Utility approach**
1. O&M expense	$2 279 000	$2 279 000	$2 270 000
2. Debt service	860 000		
3. Debt-service reserve	60 000		
4. Capital improvements	380 000	—[a]	—[a]
5. Depreciation expense		414 000	414 000
6. Taxes other than income tax			360 000
7. Income taxes			469 000
8. Return (operating income)		886 000[b]	1 451 000[b]
9. Other revenue	(79 000)	(79 000)	(31 000)
Total revenue requirements from rates	$3 500 000	$3 500 000	$4 942 000

[a] Annual cash requirements for this item are met from depreciation expense and return.
[b] Includes interest on debt.

Historical Data. In projecting revenue that may be available to the utility, it is essential to first develop adequate historical data to serve as a basis for projecting future revenues. These data, however, must be normalized or adjusted to reflect conditions that may not continue into the future. Factors affecting revenues and revenue requirements include the number of customers served, rate changes, nonrecurring sales, weather normalization, conservation and use restrictions, and price elasticity. Revenue requirements can also be affected by interest rates, changes in tax laws, and capital financing needs.

Operation and Maintenance Expenses. Operation and maintenance expenses are those incurred by the utility as it provides service to customers during a particular accounting period, and therefore do not include depreciation expenses or costs incurred in the upgrading of plant facilities beyond the level originally intended. Nonrecurring expenses, such as those generated by maintenance activities carried out less frequently than once a year, may be amortized to fit annual cost calculations. Because O&M expenses fall into a wide range of categories, it is important, particularly in larger utilities, to establish a detailed chart of accounts to classify and track the costs. A chart of accounts recommended by the National Association of Regulatory Utility Commissioners, Washington, D.C., is widely used by regulated utilities and can be modified for government-owned utilities (AWWA, 1990).

In determining revenue requirements it must be recognized that some expenses, which may appear to be O&M expenses, are actually capitalized costs. These include salaries, wages, and overhead costs on efforts devoted to capital investment projects rather than to O&M, and should be accounted accordingly.

Depreciation Expenses. Federal income tax guidelines allow greater rates of depreciation than state guidelines in determining depreciation deductions. Two systems are used for determining these rates: the class life asset depreciation range (CLADR, or ADR) system, and the accelerated cost recovery system (ACRS).

The CLADR system may be elected to determine depreciation in facilities which entered service between 1970 and 1981, and classifies depreciation costs by industry and type of asset. For facilities which entered service in 1981 or later the ACRS is mandatory. The ACRS specifies a period over which the cost of eligible depreciable property may be recovered, according to the class of property (AWWA, 1990).

Capital Expenses. Whether a utility uses the cash-needs approach or utility approach, capital expenses are a component in determining revenue requirements.

CASH-NEEDS APPROACH. Utilities using the cash-needs approach define their capital expense needs to include debt principal and interest, contributions to specific reserves, and other non-debt-financed cost-of-capital expenses, but do not include depreciation costs. While revenue requirements are not affected by depreciation, they are limited by minimum financial ratios in budgetary decisions. Some capital improvements, such as small, routine replacements, may be financed in part from annual revenues. Major projects, however, are usually debt-financed, which places the burden on future customers, who will benefit from the new or improved facilities.

In order to provide security to bondholders utilities are often required by their debt indenture to generate enough income such that net revenues (after O&M expenses are subtracted) are sufficient to exceed debt-service payments by a certain percentage. A sample utility budget is depicted in Table 8.3 (AWWA, 1990).

UTILITY APPROACH. Under the utility approach, capital-related costs include depreciation expense and return-on-rate base. Many of the factors related to the utility approach are also applicable to the regulated government-owned utility.

Annual depreciation expense may be set as a constant rate over the length of the service life of an asset or may be determined by comparing the remaining balance of the asset's original cost to be recovered against its remaining service life. In both cases the salvage value of the asset is taken into account. Using the first method the annual rate is determined as follows:

This rate, applied to the original cost of the account item, determines the annual depreciation for the asset. Estimates of average service life require extensive mathematical analysis to establish, a process which has become more sophisticated with the use of computer technology.

Table 8.3 Sample utility budget. (Reprinted from *Revenue Requirements—M35*, by permission. Copyright © 1990, American Water Works Association.)

Item	Amount, $
Annual operations	
Gross revenue	3 259 000
Less O&M expenses	2 183 000
Net revenue	1 076 000
Debt service	710 000
Net available	366 000
Nonoperating income	50 000
Total available for capital projects	416 000
Capital Project Financing	
Current capital project obligations	3 405 000
Available from annual operations	416 000
Borrowings	3 000 000
Total available	3 416 000

Because public utilities are either government-owned or franchised and have exclusive rights to operate in a particular area, they are not under pressure from competitive market forces. Since they lack the incentive to set optimal rates for their customers, utilities require some form of regulation to ensure that reasonable service at a reasonable price is provided. Rates charged by government-owned utilities are regulated by a local governing body, such as a city council or water board.

ACTIVITY-BASED COSTING. Activity-based costing is a cost-control technique that assigns costs to products or services based on their consumption of resources. It is the first step for government officials to determine exactly how many and what services are provided to citizens and the cost of those services. The ABC concept is integral to the asset management plan. The objectives of ABC (Carter, 1996) are to

- Identify component cost details, cost impacts, and savings from alternative courses of action;
- Compare delivery costs for the same service in different locations;
- Monitor present quality and delivery of core services;
- Improve efficiency of city services and customer satisfaction; and
- Achieve savings by actual reductions in existing costs by
  - Finding less expensive ways to deliver services,
  - Reducing costs of service outcomes,
  - Redesigning the program,
  - Increasing volume dependent on cost reductions, and
  - Understanding the delivery/program design interaction.

As a part of the asset management plan, ABC provides

- Useful management information;
- Objective information derived in a consistent manner;
- Measurable and increased productivity of desired services;
- Identification of business;
- Support for
  - User fees,
  - Performance measures,
  - Contracting versus in-house,
  - Program redesign, and
  - Performance budgeting; and
- Improvement in service and cost savings.

USEFUL LIFE AND DEPRECIATION. Before indicative business figures for midterm planning can be calculated, it is essential to know the useful lives of the structures. Current applicable guidelines for performing cost comparison calculations provide a starting point for determining the average useful lives for different materials. Sawatzki (1996) gives a general range for useful lives of sewers that applies to all materials.

The following procedure is one approach for how asset life is addressed for a typical utility. (Note that the information given should be used as a guide only. Individual local conditions will dictate modification.) Plants, lines, and other facilities are recorded at historical cost or, if contributed, at fair market value on the date the contribution is received. The costs of these assets are depreciated over their estimated useful lives on a straight-line basis.

Sawatzki (1996) recommends an approach to asset valuation based on classification of the condition of the sewer, taking account of factors such as the year of laying, the material, and the drainage method. Condition classes are defined by the level of structural and operational damage. Classification of a specific section of sewer is determined by inspection and is based on the largest single defect, and the number of defects, in the section. More far-reaching requirements to be met by the wastewater systems—construction in accordance with standard engineering practice and compliance with the relevant laws relating to water resources—are accounted for in the analysis with external assessment factors, such as the hydraulic conditions and the properties of the wastewater. Six condition classes are defined as follows:

- Condition class 1: damage to be repaired immediately.
- Condition class 2: damage to be repaired in the short term.
- Condition class 3: damage to be repaired in the mid term.
- Condition class 4: damage to be repaired in the long term.
- Condition class 5: damage to be repaired in the course of other construction work.
- Condition class 6: no damage.

Ordinary useful life is defined as the period in which it may be assumed with some certainty that the economic asset (in this context sewers, manholes, and pumping stations) will be fit for operation in normal service. The ordinary useful life cannot be determined exactly in advance and at present is determined only on the basis of estimated values. For cost-intensive wastewater systems in particular, the potentially large inaccuracy in the assumed ordinary useful life and thus in derived depreciation rates results in a large factor of uncertainty in asset valuation and determination of charges. The following discussion presents a statistical model that addresses the need for more accurate determination of the useful lives of sewers.

The data basis of the model comprises

- Determination of sewer condition in accordance with ATV M 143 (Sawatzki, 1996),
- Assessment of sewer condition in accordance with Sawatzki's method for determining the condition for video inspection,

- Years of construction, and
- Division of network according to service classes.

To obtain as fine a division of the collection system network as possible for the model, service functions are determined for groups of sewers having similar features, for example, section identification, type of wastewater, pipe material, and drainage method. Groups of sewers having similar properties are designated as service classes. For each service class, the sewer lengths are added up according to age and condition and converted to fractions of the service class.

The model calculates the parameters of the service functions, which represent the condition class of a sewer as a function of age. The model requires a large number of suitable data records with the decisive features. The present service time must be determined on the basis of the year of construction and the year of examination.

Calculations have shown that with the available data, meaningful subdivision of the service classes is possible only according to material and drainage method. This classification system is adequate for practical applications, that is, the valuation of assets.

Following DIN 31051 (Sawatzki, 1996), a wear reserve of 100% is assumed at the time of laying. For assessment of the condition according to Sawatzki's method, this corresponds to condition class 6, that is, a sewer without defects. Over time, the wear reserve can decrease to 0%, which corresponds to condition class 1 (see previous definitions of condition classes).

Based on an assessment of approximately 27 000 sewer sections, the model determined the average ordinary useful life to be 112 years for vitrified clay and 87 years for concrete. This assessment, however, is only a first step in improving the determination of the useful lives of sewers. It will be possible to verify the model after further sewers have been examined and the results evaluated.

To illustrate the function of the model, consider two lengths of sewer constructed in 1964 and 1944, respectively (Table 8.4). Vitrified clay with a mean ordinary useful life of 112 years was the material of construction.

The condition classification of the sewers gave a value of 3 for sewer 1. For a statistically average useful life and the corresponding reduction of the wear reserve, the condition class 4 would result for sewer 1 after 30 years. Because of the increased damage potential, the result is equivalent to that of a sewer 20 years older. Hence, by this calculation, the ordinary useful life of this sewer is now only 92 years.

On examination, sewer 2 was found to have a greater wear reserve, as might be expected from the mean useful life. This result leads to a longer useful life of 130 years.

In the case of capital goods, the useful life of which extends over several years, at the respective time of the valuation the property value must be corrected and an account taken of the useful life. The value adjustment is based

Table 8.4 Sewer line examples.

	Year of construction	Material	Age, years	Condition class	Ordinary useful life, years	Remaining useful life, years	Depreciation rate %
Example 1							
Sewer 1	1964	Vitrified clay	30		112	82	0.90
Sewer 2	1944	Vitrified clay	50		112	62	0.90
Example 2							
Sewer 1	1964	Vitrified clay	30	3	92	62	1.07
Sewer 2	1944	Vitrified clay	50	4	130	80	0.77

on a linear depreciation, in which the decrease in value of the assets is distributed proportionally over the individual periods, with an account taken of the ordinary useful life.

With the linear depreciation method, the value to be depreciated is divided up uniformly. In the event of the original useful life being altered, not only the remaining useful life but also the depreciation rate must be adapted.

These examples show the effects of the selection of different ordinary useful lives on the calculated depreciation rate and the resulting remaining useful life. The assumed depreciation rates and the associated determination of the wastewater charges are at present established for sewers having average useful lives of between 50 and 100 years, no account being taken of the structural and operating conditions.

An average useful life of 112 years can be assumed for vitrified clay. This value is used in the first example because no other information on the condition of the sewers is available. Accordingly, for sewer 1, the depreciation rate is 0.90% and the remaining useful life is 82 years.

In the second example, classifications are available for each sewer, with the result that the ordinary useful life can be calculated again. In accordance with Table 8.1, the calculations give depreciation rates of between 1.07% and 0.77% for the representative sewers. Determination of the remaining useful life gives values of 62 years for sewer 1 and 80 years for sewer 2.

The results of the model calculations show that when account is taken of the structural and operational conditions, more accurate initial assumptions can be made for determination of the ordinary useful life. The depreciation rates and remaining useful lives, which influence the assets valuation and determination of charges, are then based on the collection system sections concerned.

It should be noted that some utilities are not permitted by regulators to recover depreciation costs through their service charge rates. Instead, aggregate net debt service is used as a measure of recoverable capital cost. Consequently, expansion tends to be the driving force behind growth of the rate base. At the same time, however, the disparity between the relatively small scheduled principal payments on a growing business debt and the comparatively high depreciation and amortization costs of a mature utility's fixed assets yields significant net losses.

LIFE-CYCLE COST. Life-cycle cost is a method of calculating the total project cost, including the initial cost, annual O&M cost, and replacement costs, over the life of the project. It is an especially good method for comparing costs of projects that have high O&M or replacement costs. This method can be effectively used for comparing costs of pipeline repairs, replacements, and rehabilitation over the planning period. The repair alternative involves high annual costs, while rehabilitation or replacement involves higher initial costs and lower O&M costs.

The life-cycle cost method computes total cost by converting all costs over the planning period to present worth in a common base year or to an average annual cost. Only the present-worth method is described here. The formula for calculating present worth is presented below, followed by a short discussion of cost factors to consider when conducting a life-cycle cost analysis (LCCA).

An LCCA, sometimes referred to as a least-cost analysis, is essential to new, retrofitted, upgraded, and rehabilitated infrastructure. By definition, LCCA is a process for evaluating the total economic worth of a usable project by analyzing initial cost and discounted future costs, such as for maintenance, reconstruction, rehabilitation, restoring, and resurfacing, over the life of the project segment. The 1996 *Annual Book of ASTM Standards* contains a new standard practice (C1131) for an LCCA of concrete culvert, storm sewer, and sanitary sewer systems (ASTM, 1996).

Cost Analysis Requirements. An LCCA must be performed for all federally funded environmental protection projects. This analysis will aid in getting the project authorized and ensure that project costs are reduced for the life of the project and will reduce federal costs. It also aids in selecting procedures to rehabilitate infrastructures to their original functional condition and to strengthen or retrofit upgrade pipes to correct design or construction deficiencies and improve serviceability and functionality.

Commonly, on projects when alternate materials are bid, selection is based on first cost. However, the alternative with the lowest first cost may not be the most economical selection for the design life (material life) of the project. The relative factors that affect the analysis of two alternate materials with different service lives (material lives) for economic equivalence are first cost, project design life, material life, O&M cost, repair cost, and feasibility analysis.

First Cost. The first cost is simply the bid price for a product. This cost includes the construction cost estimate for labor, material, contingencies, supervision, and administration. Historical data may be used to determine an appropriate value for prebid evaluations. If a least-cost analysis is included as a basis for awarding a contract, then the actual bid price for the alternate materials would be used in the effective cost equations.

Project Design Life. Service life is defined as the number of years of relatively maintenance-free performance. The project service life for culverts and sewers under primary and secondary roads is a minimum of 50 years, and these structures should be designed for 100 years of relatively maintenance-free performance, using an appropriate safety factor. Note that a high level of maintenance may justify replacement before a failure occurs.

Material Life. Agencies such as the U.S. Federal Highway Administration,

U.S. Army Corps of Engineers, Soil Conservation Service, and most state departments of transportation have published reports for predicting the material service life of sewers. Historical data, pH values of surrounding soils, water, potential for abrasion, bed load parameters, and ultimate pipe slope form the basis of predictions.

Operation and Maintenance Costs. These costs include the typical O&M costs of the structure plus replacement, if necessary. There are several types of maintenance that will extend the service life of sewers:

- Routine maintenance—to keep sewers in a uniformly good and safe condition by repairing specific defects as they occur. Routine maintenance is often "reactive" as opposed to "scheduled." This includes debris or sediment removal in response to heavy rainfall.
- Preventive maintenance—cyclic in nature and can be scheduled to prevent progressive deterioration. Typical activities include scour prevention, joint sealing, mortar repair, and cleaning.
- Rehabilitation—most appropriate for sewers that have deteriorated beyond the point of effective maintenance but have not deteriorated to the point where all structural integrity has been lost.
- Upgrade to equal replacement—to provide service that is equal to that provided by a new structure. Upgrading may include such items as lining of the barrel and provisions for safety improvements.
- Replacement—to provide a completely new structure with a new service life; applicable when the condition of the structure has deteriorated to the point that there is little structural value left.

Repair Costs. Repair costs include the scheduled and unscheduled costs to repair the structure.

Feasibility Analysis. All projects must be viable for at least 50 years. The rate of inflation and the discount rate for money will control the feasibility of most projects. Therefore, a present-worth analysis of the project is calculated using a value called the inflation interest factor (IIF). Assuming that inflation and discount rates show that a good value for IIF is 0.98, present worth is calculated as follows:

$$\text{Present worth} = \text{Future cost } (0.98)n \tag{8.1}$$

where n is the number of years to that future cost. For example, the future cost will be \$150 (for 20 years into the future) if the present worth is \$100.

Conduit Product Life Data. Each application has its unique set of conditions depending on soil conditions, groundwater, placement, and other factors. The following are sample product lives; manufacturers should be contacted for the specific application:

- Concrete—most studies indicate a product life of 70 to 100 years. The National Association of Sewer Service Companies (NASSCO, 1995) recommends 75 years.
- Steel—there are various coatings, such as bituminous coatings, metallic coatings, fiber bonded coatings, and concrete linings, that can be applied to corrugated metal pipe (CMP) to extend the life to 50 years (NASSCO recommends 25 to 75 years).
- Ductile iron pipe—100+ years (225 utilities have cast iron pipe more than 100 years old; NASSCO recommends 75 years).
- Plastic—when a plastic pipe is well placed in firm, compacted soil and not exposed to ultraviolet radiation, the plastic pipe may have a design life of 50 years (as recommended by NASSCO). The installation histories of these pipes are short; therefore, their performance life is difficult to predict.
- Clay—more than 100 years (Clay Pipe Engineer Manual) (NASSCO, 75 years).
- Fiber glass—more than 100 years (NASSCO, 50+ years).
- Polyethylene—more than 100 years (NASSCO, 50+ years).
- Cured-in-place pipe (NASSCO, 50+ years).
- Fold and formed pipe, 50+ years WSF2.
- Chemical grout—NASSCO recommends 20 years.

Construction Cost and Economics Analysis. Economy is essential if progress is to be made in dealing with infrastructure problems, but the emphasis on economy must be tempered or moderated with the need to ensure that safety and long-range costs are not compromised.

There are several types of maintenance and repair that will extend the service life of sewer pipe. Each of these maintenance types has different cost implications that must be borne by the agency and the service that will be received by the public. These alternatives should be carefully documented, and the cost should be analyzed so that the best value can be received.

Procedure for Determining the Life-Cycle Cost Analysis. First, establish the specific objectives of a project and identify alternative ways of accomplishing the objectives. Next, establish basic criteria for project design life (material or structure service life), direct and indirect costs, and time of maintenance, rehabilitation, and replacement. Compile data required to compute the LCCAs of potential alternatives. The costs of planning, design, engineering, construction, maintenance, rehabilitation, and replacement are adjusted for inflation, residual value, and the time periods for all future costs. Compute the LCCA as follows:

$$\text{LCCA of a material or structure} = \text{Original cost} - \text{Residual values} + \text{Cost of maintenance} + \text{Rehabilitation} + \text{Replacement} \quad (8.2)$$

STAFFING. The actual task of getting accurate, reliable information from collection systems is unique. With the hardware and software currently available, workers bring increasingly specialized professional skills to the operations environment. The challenge is to build quality control and verification into staff performance.

Accurate sewer performance information is one of the core tasks of an asset management program, and the first step is to ensure that each worker knows his or her role in the entire process. This team member approach should be invoked repeatedly throughout the training process so that the link between individual decisions in the field and the overall result is clear. The link to pollution remediation and asset protection should be established early and emphasized often.

The second step is to change the focus of training from machinery operation to monitoring of collection systems conditions. This approach is particularly useful in developing workers who will specialize in producing accurate information over several years. Technology will change, but the basis for using technology and the keys to quality will already be present in workers regardless of the technology they are using.

The advantages of this approach have been appreciable. First, based on past experience, collection systems field workers are more likely to make independent decisions affecting quality on site. Typically, collection systems personnel work outside direct professional supervision most of the time. Review by professionals or supervisors may provide quality control, but it is largely after the fact. Second, even if workers do not have all the answers, they understand enough of the goals to raise issues and questions at the right time. Third, workers are motivated to use their specialized training. If the challenge is to incorporate quality up front, sewer workers must be able to make informed, correct decisions on their own.

EXISTING COLLECTION SYSTEM APPURTENANCES. There are three critical components of any type of asset

- Inventory,
- Condition, and
- Performance.

Most cities have a good handle on the first two items, but the third has been proven to be elusive because it has been difficult to provide tools of even moderate accuracy that are sufficient to measure performance. A collection system performance indicator data collection form will help overcome this limitation. Typical information to be shown on this form includes

- General information (name, address);
- Collection system description (miles of sewer population);
- Special conditions (local conditions, problems);
- Age distribution of collection system;
- Size distribution of collection system;

- Distribution of gravity sewer by material;
- Distribution of force mains by material;
- Preventive maintenance of system, including
  - — Physical inspecting of a collection system,
  - — Preventive maintenance, and
  - — Preventive maintenance cleaning;
- Stoppages;
- Repairs and rehabilitation, proactive;
- Repairs and rehabilitation, reactive;
- Pumping stations (horsepower, number of pumps, remote status);
- Pumping station failures;
- Force mains (inspection, monitoring, failures);
- Air-relief and vacuum valves (inspection, failures);
- System O&M efficiency (O&M hours, labor downtime);
- Level of service (rate, complaints, public health, claims);
- Financial (rate, asset value, O&M cost, capital improvement plan [CIP], personnel training);
- Safety (hours of assigned O&M, lost time);
- Regulatory (violations, fines, wastewater treatment plant [WWTP] upsets);
- General (sanitary sewer evaluation study [SSES], GIS, CMMS); and
- Procedure or other available documentation (overflows [OF], cleanup, budget, long range CIP, grease, pretreat, private inflow and infiltration [I/I]).

There may be some unique features about a system that should be added to this list to allow an adequate understanding of the system. Additional information on performance indicators is included at the end of this chapter.

CUSTOMERS. The wastewater collection systems manager is faced with a mandate from the time the sewer system is first constructed: to provide reliable service and customer satisfaction, often with less-than-adequate resources. The lack of reliable service may be measured by increased customer complaints, property damage resulting from backup into homes, regulatory noncompliance orders and fines, sanitary sewer overflows (SSOs), and political pressure to solve the problem. Sewer use fees are designed to provide a level of service that precludes raw wastewater overflows and provides customer satisfaction.

FUNDING AND GRANTS. Water utility managers establishing or changing rates must develop pricing structures that will cover operating costs and future expansions, provide stable and predictable revenues from year to year, and be equitable to all customers. Rate structures also should be easy to understand and implement, should promote water conservation, and should meet local, state, and federal regulations. The unique characteristics of each service community, such as climate, geography, service area, and any general fund subsidies or grant assistance, should also be taken into account.

When these criteria conflict, those who are assigned the task of setting rates must prioritize their goals and make compromises. An asset management approach will help meet these goals.

Since 1972, billions of dollars have been spent at all levels of government in the U.S. to meet the requirements of the Clean Water Act. Unfortunately, little of the U.S. Environmental Protection Agency's (U.S. EPA's) construction grants has gone toward rehabilitation or replacement of the country's 19 000 collection systems. These grants were primarily focused on expansion of WWTPs (Public Law 92-500 in 1972). The 1987 amendments to the Clean Water Act brought more emphasis on sewer rehabilitation to reduce the hydraulic loads on WWTPs from extensive I/I. This emphasis lowered capital costs associated with oversized facilities, reduced O&M costs, and prolonged the life capacity of the treatment and conveyance facilities. Asset management takes this one step further by predicting O&M costs more efficiently.

Increasingly, states are requiring grant applicants to determine the I/I or SSES conditions in the sewer systems before applying to expand or upgrade a WWTP (U.S. EPA, 1991). There are several funding sources that provide low-cost loans or grants for this type of work:

- Community development block grants—these grants have been used by local units of government to fund planning studies, sewer rehabilitation, and construction of collection systems. Typically, projects must benefit 51% low-to-moderate income facilities. Projects must demonstrate need.
- U.S. Department of Commerce, Economic Development Administration public works and economic development—these grants have been available through state or federal agencies to counties, cities, towns, and regional sewer districts to economically distressed communities. Projects typically must be related to creating or retaining jobs.
- State revolving fund—these low-interest loans are available to political subdivisions including cities, towns, counties, and regional sewer conservancy and sanitary districts. Projects must be related to abating pollution associated with existing problems. The 1996 Clean Water Act State Revolving Fund also provides hardship grants to communities of fewer than 3 000 residents.
- Rural economic and community development services—low-interest wastewater loans and grants are available for public and nonprofit organizations serving fewer than 10 000 residents when other financing sources are not reasonably cost effective. This funding is primarily used for rural residents. Eligible communities must demonstrate need for the project.
- State programs—many states have set up competitive programs for

grants and loans. Information for these programs or any of the preceding programs can be found at most regional U.S. EPA offices.

In most cases, grants and loans of these types are subject to specific funding cycles; time schedules; wage rates; legal, financial, and administrative reviews; land acquisition; eligible cost; and grant match. The effect of these requirements and conditions should be carefully studied to determine the best funding for each project.

Sewer grants are becoming more competitive and harder to obtain. Whereas in 1975 projects cost $1.00 and federal grants were paying 75% of the costs, local funds only paid $0.25 on the dollar. In 1995, costs were 2.5 times those in 1975. Escalating costs (a $2.50 ÷ 0.25 factor of 10 times) have caused a reassessment of the need for new construction.

FLEET MANAGEMENT. Chapter 3 provides detailed information on vehicle and equipment requirements for collection systems. Effective fleet management requires a comprehensive plan to address the numerous aspects of the fleet's operation, such as budget issues, labor requirements, recordkeeping, and the scheduling of vehicle maintenance and replacement, and to ensure that state and federal regulations, such as clean air requirements, are met. This section outlines a strategy to optimize these assets and operating costs.

With careful planning seasonal maintenance scheduling should be conveniently coordinated with the peak and off-peak operating seasons, and the workload of the fleet management staff should be well distributed from one day to the next. Managers should have a good picture of the overall fleet budget as well as the costs of individual operations. Ownership or leasing of vehicles and other equipment should be optimized, and the day-to-day activities of the employees should be appropriate to their expertise. Over- and understocking of essential parts should be avoided. In general, a strategic rather than a "keep it afloat" approach to fleet management will allow the utility to avoid unnecessary expenses.

A vehicle and fleet operation consists of five main components:

- Inventory (vehicles, equipment, and parts);
- Labor;
- Facilities;
- Budget, purchasing, and support services; and
- Recordkeeping.

To implement a strategic fleet management (SFM) program, the following steps are needed (Taylor, 1995):

- Plan and develop a budget, establish a working group, and develop a time line;

- Develop objectives and goals for the fleet;
- Create measurable operational definitions of the objectives;
- Identify and choose among alternative courses of action;
- Choose the courses of action and activities necessary to reach the objectives and goals;
- Examine whether the organization has the expertise to effect change;
- Create an environment and organizational structure that is ready for change (meet with management and line staff to motivate them for change);
- Collect and analyze baseline information;
- Develop and implement programs and procedures—inventory programs, labor programs, and facility enhancements; budget, purchasing, and support service policies; and recordkeeping systems;
- Evaluate organizational changes and programs; and
- Continuously monitor operations.

Before an SFM program is set up it should be determined whether the fleet management department has the time and expertise to develop the program in-house, or if a consultant should be hired to direct the development and implementation. Once the program is in place it may be evaluated using the following checklist (Taylor, 1995):

- Inventory:
  - — Evaluate fleet specifications,
  - — Conduct a detailed physical examination,
  - — Analyze vehicle and equipment use,
  - — Review and analyze inventory system,
  - — Install simplified parts numbering system,
  - — Introduce stocking levels,
  - — Improve availability of parts, and
  - — Reduce inventory losses through security controls.
- Labor:
  - — Schedule the preventive maintenance system,
  - — Analyze labor,
  - — Conduct employee performance analysis,
  - — Schedule labor,
  - — Troubleshoot equipment, and
  - — Provide training.
- Facilities:
  - — Review garage facility,
  - — Examine fueling, and
  - — Ensure compliance with regulatory requirements.
- Budget, purchasing, and support services:
  - — Prepare annual fleet budget,
  - — Prepare fleet capital budget,

- Maintain vehicle and equipment replacement schedule,
  - Practice public bidding and buying smarter,
  - Recapture more warranty,
  - Alternate fuel vehicles, and
  - Examine privatization.
- Recordkeeping:
  - Perform general recordkeeping,
  - Practice computerization,
  - Maintain daily vehicle operator reports,
  - Practice expense budget recordkeeping, and
  - Maintain an accident review system.

Note that many of the elements of the SFM program are also applicable to the collection system (gravity sewers, pumping station, force mains, and manholes).

SETTING FUTURE GOALS

Physical inspection of collection systems remains the foundation of an effective O&M and asset management program. As a labor-intensive process, an effective collection systems O&M program depends on the willingness of utility managers and elected officials to recognize that reactive maintenance often exceeds the costs of effective and proactive O&M programs and of protecting the value of the city's largest asset.

An increasing number of municipalities, utilities, cities, and towns are evaluating the worth of their infrastructures. While most publicly traded utility companies have comprehensive financial balance sheets that depict historical costs for their property, plant, and equipment (including annual and accumulated depreciation), the majority only capture this information as general ledger totals, with amounts added or subtracted based on capital expenditures. Billions of dollars in assets on the financial ledgers of utility companies cannot be traced to the field where they are located.

There are several reasons that this is a problem. Currently, the main reason is that it is not possible to accurately predict degradation of sewers caused by aging or corrosion. Research currently being conducted will explore predictive techniques to enable planners to establish optimum maintenance schedules for repairs or replacement. Accordingly, factors that adversely affect sewer life need to be identified.

As previously mentioned, physical inspections of collection systems are the foundation of the asset management system. U.S. EPA and state officials constantly remind managers to know their own systems. This practice should continue so that the collection system manager will be ready to begin true asset management of the system when all the tools are available.

Data gathering during field investigation and hydraulic modeling should continue, allowing for the prioritization of system needs and establishment of the rehabilitation options. *Existing Sewer Evaluation and Rehabilitation* (WEF, 1994) has extensive information on this subject.

Although the cost of construction has increased, the cost of I/I removal has gone down. The cost of rehabilitation effectiveness has increased because of new technology to locate and diagnose problems. Typically, I/I removal and subsequent rehabilitation are becoming increasingly cost effective. In the future, to be successful, engineers will be required to be rehabilitators, optimizers, and diagnosers.

ATTAINING FUTURE GOALS

Public works and utilities are adopting asset-valuation-based infrastructure management systems to obtain both the data and financial relationship of assets that support their customers. By recognizing the annual revenue received from each customer and capturing the operating expenses of the infrastructure delivery system, public works, utilities, and municipalities can value their asset base and perform customer profitability. Two tools to accomplish this are return-on-investment and return-on-assets analyses.

When accurate collection systems performance indicators and predictive tools for degradation of sewers from aging or corrosion are available, asset management programs can be developed. The extension in the useful life of the collection system by means of replacement or rehabilitation can be predicted so that adequate funding is available to protect this large asset. Management will also have to become proactive in implementing a preventive maintenance program. In the future, this program will develop into an asset management program based on predictive maintenance and performance indicators.

Management must provide adequate funding for staff training. This factor will become more important as information technology has an increased role in collection systems management. A qualified collection system employee will become the most critical element of a maintenance program.

Infrastructure is fast becoming "infostructure" (Ward, 1994). Information systems will play a key role in inventory, analysis, and public access for the collection system. Geographic information system technology is emerging as a key tool in making decisions in a cost-effective manner. This technology is causing many different utilities to look at GIS together as one group. Sewer, water, gas, cable television, electric, streets, and property evaluation have all joined together in an effort to pay for the GIS system. A benefit of this practice has been that with a common database, utilities can plan together "in real time." These several databases may now be found in one system at great savings. In the past, it was not uncommon for the street department to pave a street one month and the next month for a sewer or water department to dig it up. Sharing a similar GIS database can put these different functions together and eliminate or reduce the chances of these kinds of practices occurring.

Public awareness and support for long-term maintenance and rehabilitation are crucial to the success of the asset management program. One way to accomplish this support is through public meetings. Feedback from individual customers can assist in providing information on backups. This will help in analyzing the problem and getting the customer involved in the solution. Backup information such as when, how much, where, how long, how high, and damages incurred provide the city with an insight into prioritizing investigation efforts.

ACTIVITY-BASED COSTING. Components. There are four basic components that determine how the ABC model will track cost information: activities, cost pools, drivers, and outputs. *Activities* represent the services performed or products produced by an organization. The goal is to determine the total cost of the activities as organized along the respective cost pools. Examples of city or utility activities include cleaning sewers, inspecting residences, removing solid waste, patching chuckholes, and issuing permits.

Cost pools are the main components of costs incurred by an organization. Activity-based costing allocates cost pool amounts to activities. Examples of cost pools include types of labor, types of vehicles, types of direct materials, and types and amounts of overhead allocation.

The method used to allocate costs from the cost pools to the activities is defined through *drivers*. When drivers are defined, the user must be capable of establishing the total quantity of the driver. Drivers are then linked to cost pools to determine the rate per each driver unit. Examples of drivers include direct labor hours, equipment hours, and material quantities.

Output is the unit of measure that represents the number of times an activity is performed during a specified period. Examples of outputs include miles of sewers cleaned, number of inspections performed, number of chuckholes patched, and number of permits issued.

Table 8.5 (Carter, 1996) is an example of solid waste ABC that can easily be applied to a heavy or light cleaning of a sewer. This information can be used to determine costs for specific tasks or to determine whether to outsource specific tasks.

With increased focus being directed to sewers, more and more emphasis will be put on maintenance and repair of sewers. Some sewers have not been looked at in more than 50 years. Inventories of the system will become periodic in an effort to reduce excessive flow induced by rain. This process is labor intensive.

In the transition period from reactive preventive maintenance programs to more proactive preventive maintenance, a peak staffing period may occur. The use of in-house staff versus a contracted workforce versus a mix of both methods of staffing should be considered. As an example, various parts of the inventory and evaluation (such as cleaning and television or flow monitoring) may be appropriate to contract out.

Table 8.5 Activity-based costing; solid waste example (Carter, 1996).

Cost Pools		Cost Pools and Related Drivers	
Cost pool name	**Total dollars**	**Cost pool name**	**Driver**
Refuse driver salary	$2 000	Refuse driver salary	Driver hours
Refuse collector salary	$1 800	Refuse collector salary	Collector Hours
25 yd truck costs	$ 600	25 yd truck costs	25 yd truck hours
Central service overhead	$ 400	Central service overhead	Total labor hours
Miscellaneous overheads	$ 300	Miscellaneous overheads	Total labor hours
Total	$5 100		

	Activity Consumption		
Driver	**Residential trash**	**Heavy trash**	**Total**
Driver hours	50	50	100
Collector hours	200	100	300
25 yd truck hours	25	15	40
Total labor hours	75	65	140

	Cost per Activity		
Driver	**Residential trash**	**Heavy trash**	**Total**
Refuse driver salary	$1 000	$1 000	$2 000
Refuse collector salary	$1 200	$ 600	$1 800
25 yd truck costs	$ 300	$ 225	$ 600
Central service overhead	$ 215	$ 185	$ 400
Miscellaneous overheads	$ 160	$ 140	$ 300
Total	$2 950	$2 150	$5 100
Outputs:			
Tons—30	$98.33	$ 71.67	
Homes—200	$14.75	$10.75	

CAPITAL PROGRAM PLANNING. The legal context under which municipal utility decisions have historically been assessed is that of utility law. The standard holds municipalities to a duty similar to the one imposed on a privately owned public utility, that is, that service be supplied to all customers in similar positions to those already being served. Courts analyze municipal decisions relating to extensions of service on the basis of reasonableness and whether or not the extension discriminates against any individual or group of property owners.

While utility law currently is the common law rule applied in most courts, new legal emphasis is being placed on municipal planning for growth and regional coordination of planning. Several states have complex legislation that links capital improvement programming directly to comprehensive plans. Some courts have recognized the value that comprehensive planning considerations have to utility decisions. Whether or not a community is under the stress of rapid growth, the incentive to plan for future growth is present in nearly all of the states.

A capital improvement program provides an excellent basis for making the link between growth management plans and utility policy. It allows a community to plan for future development by realistically assessing utility obligations, while formulating its utility policy to reflect the community's goals toward growth.

An integrated capital improvement program should carefully plan financial policy for extension and expansion of sewer systems. An increasing number of states are allowing the costs of system growth to be placed on new users. As financing of new service and system rehabilitation becomes a budgetary burden on localities, the capital improvement program can provide a sound basis for equitably and legally distributing the costs of system expansion.

In light of continuing changes in the legal importance of planning for growth, it is clear that state courts will have to address the conflict between utility law standards and the increasing need for municipal planning and growth management. When dealing with a challenged municipal utility policy, courts will have to determine a basis for that policy and may run up against conflicting objectives for sewer and water provisions. Objectives must balance growth management principles with what can be legally recognized as "utility-related" rationale. If such a policy is clearly outlined by a capital improvement program that integrates utility cost and revenue projections with local growth management objectives and land use design, a reasonable basis for judicial sanction will become clear.

Utility providers will not be able to operate in a vacuum. As service provision becomes more expensive and more important to metropolitan growth, an integration of values must form the basis for future utility extensions, rehabilitation, and financial policy (University of North Carolina, 1987).

GRANTS. Current federal grant monies are dwindling and have become more competitive. Managers new to the system should examine other successful projects, such as requesting from a lending or grant institution information about successful applicants and the team they used. These teams typically include engineers, attorneys, and financial consultants. Imitating the strategy of a successful team may be beneficial.

To implement the project successfully, the funding plan must be affordable. Because financial resources vary considerably from one municipality to another, there are no established criteria for an affordable project.

Affordability will depend on existing debt obligations and operating expenses for other public works in the community. User fees may already be too high, hindering the ability to move forward with any new improvements.

Grants and loans are often more readily available in poorer (below median household income [MHI]) areas (OhioScoop, 1997). The Ohio Small Communities Cities Community Development Block Grant Program, for example, requires rates at a certain level to be considered. In this case they require that combined water and sewer fees be more than 2.5% of area MHI before consideration. Similarly, new sanitary sewer system fees must be at least 1.5% of the area MHI.

The general approach used for developing a funding plan is to determine projected user fees under several different funding scenarios spanning the expected combination of grants, loans, bonds, cash reserves, and other sources of capital. If the overall project is too large to implement all at one time, phasing can be considered.

MANAGEMENT INFORMATION SYSTEMS. As utilities convert to a computerized network to handle schedules, plans, specifications, engineering drawings, contracts, change orders, and time sheets, project communications and management of projects are headed toward a paperless management system. Computerized systems will allow managers more time to manage projects; better and faster communications; less time and money spent producing and reproducing reports, plans, and specifications; and a reduction in clerical staff.

The recognition that businesses organized by departments and work functions are having difficulty achieving outcomes and appropriate levels of customer service has created a high degree of interest in the concept of reengineering organizations (Finkle, 1995). With increasing frequency, organizations are undertaking enormous soul-searching efforts in an attempt to define the organization's mission. In some cases, disruptive reorganizations that attempt to reshape and reengineer the workforce follow.

In more cases, however, the fundamental issue is how to design or adapt the organization's information systems to serve both the specialized information needs of specific work functions and the organization's requirements for sharing and integrating information. For water and sewer departments (as well as other organizations focused on infrastructure management), this often has less to do with reengineering organizational structure and more to do with reengineering the information systems that support the organization. Because approximately 75 to 80% of the organization's information relates to location, the proper implementation of automated mapping/facilities management (AM/FM) or GIS technology can result in a dramatic reengineering of the organization's ability to effectively use information (see Figure 8.2). The following example illustrates many of the information management issues inherent in most sewer and water utilities.

When the city of Wichita, Kansas, began its AM/FM project, it started with a comprehensive needs assessment—an assessment that broadly examined the utility's information needs for support of its critical work processes. The goal was to determine what the information system should look like to produce faster, more efficient, and smarter work. To begin the analysis, however, a careful examination had to be conducted to assess how the utility was currently managing information and what issues and information system dysfunctions existed that needed to be overcome.

As the analysis proceeded, substantially more information system dysfunctions were key impediments to management's ability to effect funda-

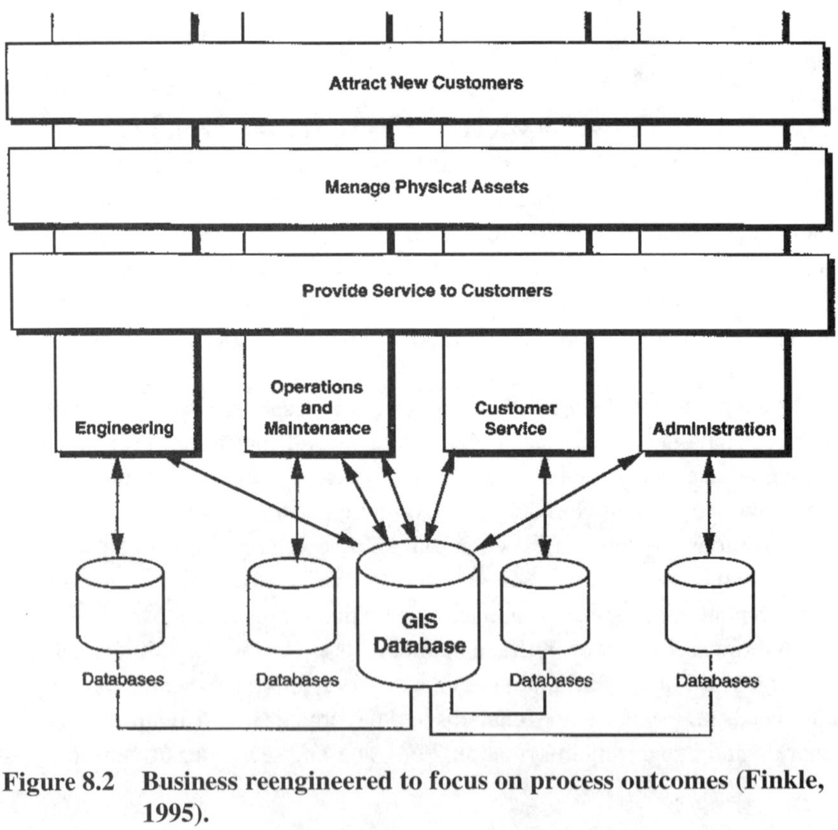

Figure 8.2 Business reengineered to focus on process outcomes (Finkle, 1995).

mental and lasting operational improvements. As with many sewer and water operations, maintenance work activities were largely dominated by responding to emergencies, with scheduled maintenance receiving substantially less attention. Though not the sole cause, limitations in the department's ability to manage and manipulate information directly contributed to the problem.

A primary example was the utility's limited ability to analyze information. While complaints and closed-out service orders were carefully filed, the paper-based system being used provided maintenance supervisors with no real capacity to analyze the repair events that were occurring. Trends could not be easily determined, and the clustering or randomness of events was difficult to discern. Without the ability to statistically manipulate and correlate complaints and repair histories, maintenance supervisors were extremely disadvantaged in their ability to anticipate and correct problems.

Multiple work order generation and tracking systems that relied on a mix of automated and manual systems were in use. As a result, the work order processing system did not provide a consolidated "cradle-to-grave" tracking system. Documentation of scheduled work was not adequately linked with documentation of completed work; thus, the department's ability to manage workforce productivity and its ability to track and report work backlogs and the completion of scheduled work were impaired.

Another problem being experienced was that the predominate focus of information collection was on documenting activity, with little information being collected to support measurement of productivity. For example, information on the kilometres (miles) of pipe cleaned or the number of complaints responded to was recorded. What was not being collected, however, was information that would allow the calculation of labor effort expended per kilometre (mile) of pipe or the time it took to respond to the complaint. As sewer and water departments struggle to perform more work while reducing cost, information of this type will be critical for measuring and improving workforce productivity.

After establishing an understanding of its information system issues and dysfunction, the utility was able to define an approach for improving its operating efficiencies and effectiveness. The approach adopted relies predominately on reengineering of existing information systems using AM/FM/GIS technology as an information systems integrator. The information system development program underway at the utility is oriented to achieving three basic system objectives:

- Implementation of an integrated facility asset recording system,
- Development of an integrated work-order processing system, and
- Development of a "single point of entry" addressing system.

To directly improve efficiencies in its management of maps and records related to facility assets, AM/FM/GIS technology that is integrated with maintenance management software (MMS) was implemented. By using an integrated system, field changes will be captured and recorded to the facility database and maps as field crews record their activity on work orders, and this information is entered to the shared AM/FM/GIS–MMS database when the work order is closed out.

The second component of the utility's information system integration strategy is to consolidate work order processing and status reporting. With implementation of MMS, consistent work order generation and reporting will be established. To achieve an optimal level of consolidation, however, data sharing linkages are planned to be implemented between the customer billing system and the MMS. With this link established, the utility's telephone center representatives will be able to access information on the status of both service orders and work orders, providing them with an enhanced ability to respond to citizen inquiries. In addition, with automation and integration of work orders and linkage of the MMS with the AM/FM/GIS, maintenance supervisors will have the ability to statistically analyze complaint and repair data. Applications are being developed that will allow the maintenance planner to construct custom queries and to prepare pin maps and other thematic maps that show the type, frequency, and clustering of events.

The third component of the utility's information system integration strategy is to reduce redundancy and information conflicts by implementing data sharing between its two address databases—the building and sewer permit system and the customer billing system. When this planned integration is complete, the customer billing system will receive the addresses entered into the permit system when a sewer account is established, eliminating the need to reenter an address to establish a water account.

In a computerized management system, the Internet will become an increasingly more important source of information. Discussion groups and bulletin boards are available for individual managers to discuss and solve problems. The latest U.S. EPA technical data and regulations and industry technical manuals will be available on the Internet to speed information dissemination.

INTEGRATED INFRASTRUCTURE MANAGEMENT SYSTEM. Infrastructure, an interconnected system of roads, sewers, public buildings, power plants, parks, and other major facilities, is crucial to environmental quality, economic well-being, and quality of life in general. Clean water, effective waste removal, reliable energy supplies, transportation, and communication have supported growth and prosperity for a steadily growing proportion of the U.S. population. However, aging facilities, networks that no longer fit well the patterns of demand, and recurring conflict among diverse governmental and private sector agencies now pose problems. Innovation and integrated infrastructure management offer the promise of higher returns on the public's infrastructure assets.

Purdue University (West Lafayette, Indiana) and the city of Indianapolis, Indiana, formed the Infrastructure Innovation (I^2) Partnership in 1995 under the overall leadership of Purdue's School of Civil Engineering. The I^2 Partnership is undertaking an integrated program of research, education, and technology transfer, designed to produce and disseminate new information that enables enhanced infrastructure performance and to encourage application of this new information so that infrastructure performance is improved.

At the core of the program is the concept of a new management tool for infrastructure system administrators, the Integrated Infrastructure Management System (IIMS), a computer-based management tool that will apply advanced information collection and management technologies to provide more efficient, accurate, and effective bases for making decisions about infrastructure. The IIMS will combine inventory, condition assessment, predictive modeling, scenario development, and user-friendly information access capabilities.

The IIMS will be designed to support informed decision making about development, operations, and retirement or redeployment of infrastructure assets. The IIMS will be a program-level tool, intended to support broad budget and policy decisions. The IIMS database and analysis procedures will draw on and be compatible with the more finely detailed project-level information that is used, for example, by city design engineers.

Infrastructure is a public asset intended to provide support and service to a wide range of social and economic activities. Integrated infrastructure management seeks to achieve high returns on the public assets that infrastructure represents. The measures of infrastructure performance should be indicative of economic, environmental, and social outcomes of providing infrastructure services.

An essential problem facing infrastructure decision makers is budgeting overall levels of spending among functions (such as expansion, repair, maintenance, and renewal) and modes (such as roads, sewers, and drainage). The I^2 Partnership's program will seek to ensure that the IIMS, as a management tool, provides information to assist in making effective budget allocations within this context of complex constraints.

To make their decisions, the community and its leaders must judge whether the infrastructure is performing adequately. The judgment is made not only by users, but also the utility's neighbors, businesses and households that depend on the infrastructure's services, and the taxpayers and government officials who own and operate the system. Measuring the condition of a region's infrastructure depends on being able to measure performance as well. Devising effective ways to measure and monitor infrastructure condition and performance is another important part of the program.

The IIMS will have several principal components (See Figure 8.3). While substantial research will be required to resolve the issues facing the I^2 Partnership's program, many existing procedures, tools, and resources can be used, at least initially, in crafting the IIMS. The city of Indianapolis, for example, has recent inventories of sewers, stormwater culverts, and bridges within its jurisdiction and uses a pavement management system that incorpo-

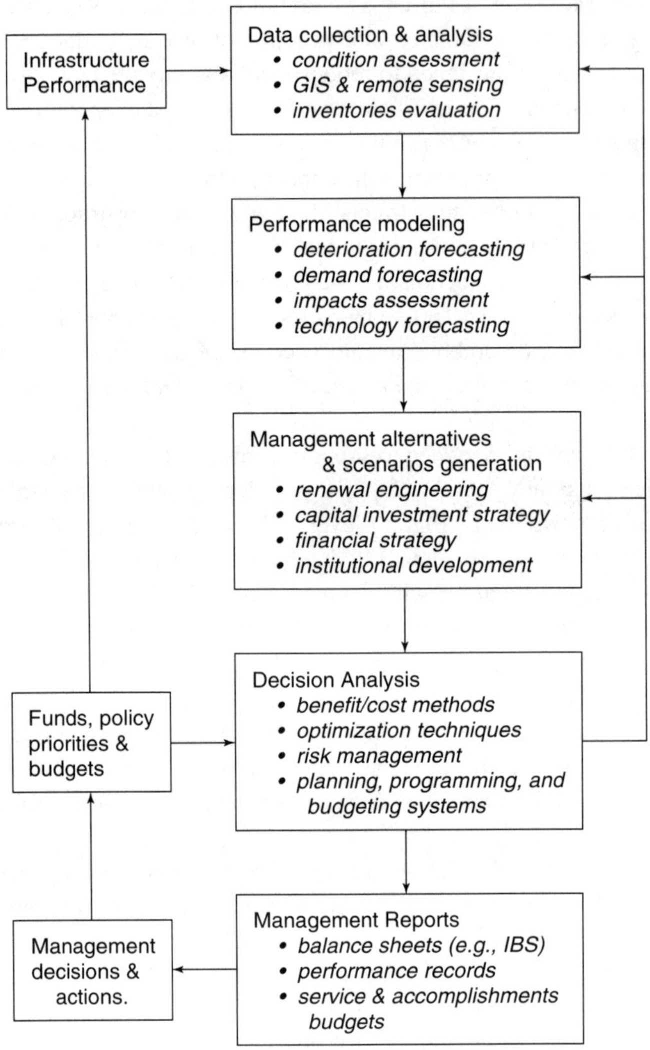

Figure 8.3 Conceptual structure of the IIMS.

rates a detailed inventory of highway and street pavement conditions. These various inventories will become part of the IIMS implementation in Indianapolis.

Additional work may be required to integrate data on these several subsystems into a single cadastral framework to produce a fully developed GIS. Other local or state governments joining the I^2 Partnership may have different data collection and analysis resources.

The School of Civil Engineering and Krannert School of Management at Purdue, with the city of Indianapolis, initiated the I^2 Partnership. The Indiana Economic Development Council, Indiana Department of Transportation, Consulting Engineers of Indiana, Indianapolis Chamber of Commerce, Indiana Constructors, Inc., and the MATRIX group, Inc., joined in early definition of the I^2 Partnership program. The partnership members plan to seek participation by new industries, government agencies, and education institutions. Preliminary linkages have been established with Greensboro, North Carolina, through the participation of North Carolina A&T University, and with Baltimore, Maryland, through participation of the MATRIX Group, Inc. During the next few years, the I^2 Partnership will focus on research for developing better technical (that is, deterioration, impact assessment, and technology forecasting) and valuation (that is, financial, economic, social, and environmental) models that can be used in an initial formulation of an IIMS for the city of Indianapolis.

The I^2 Partnership envisions further that a central feature of the IIMS may be an infrastructure balance sheet (IBS) (see Figure 8.4), a management report aimed at giving decision makers a realistic assessment of the value of infrastructure. Analogous to a corporate balance sheet, the IBS will represent infrastructure as both assets and liabilities that a region (for example, a municipality or an urban region) develops and uses to support the economic and social well-being of the people who live there. Like the corporate balance sheet, the IBS may be presented in succinct form but embody a wealth of more detailed information (for example, in the footnotes of the corporate re-

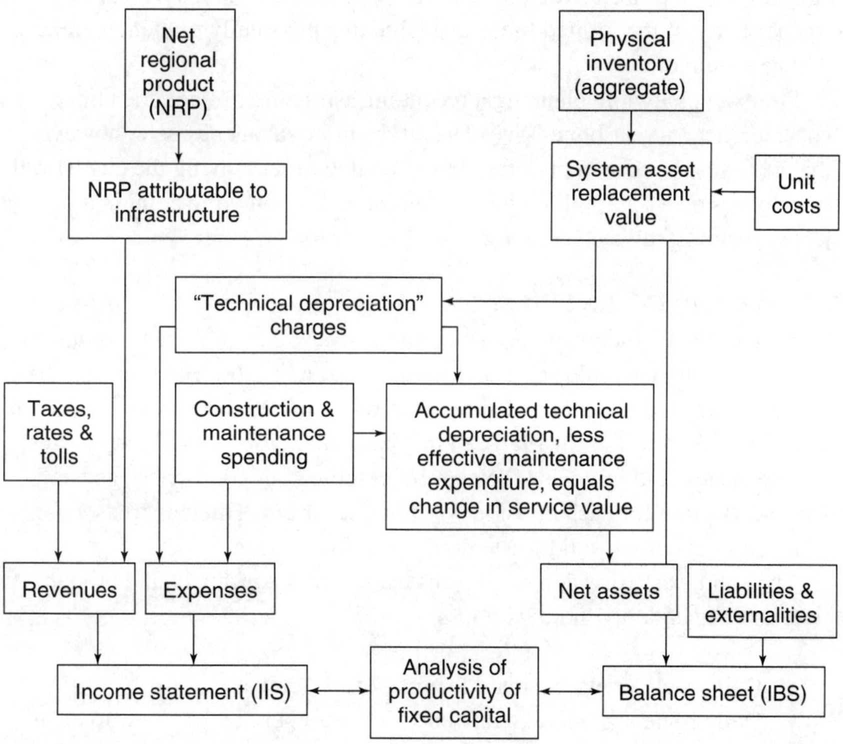

Figure 8.4 The IIMS development accounting structure.

port and other supporting documentation and analyses) for those who have a use for such detail.

Current plans are that members of the I^2 Partnership will be linked electronically, using the resources of the Internet, to establish an experimental "virtual" research institute. While participants located in Indianapolis, West Lafayette, and other cities may be separated by substantial distances, within the virtual institute they will be able to maintain close collegial relationships with the frequent and casual exchange of ideas. The network links may include the capability for video teleconferencing and other means for encouraging group activities without incurring high travel costs. This experiment is one of the key elements of the I^2 Partnership's explorations of what may be the future of U.S. and global infrastructure.

POLITICAL INTERFACE. The term *unfunded mandate* has been used in an attempt to reverse the regulatory pressure on municipalities to provide funding for their aging sewers. Unfunded mandates exist within the myriad policies cultivated by local, state, and federal government. However, the SSO issue does not fall within this realm. Sewer use fees are designed to provide a level of service that precludes raw wastewater overflows and provides customer satisfaction. This *funded* endeavor has been inadequate for many municipal public works departments, with the resultant deterioration of the collection system infrastructure. To bring collection system assets back to standard, use of the limited funds available in a physically prudent manner will be required.

For example, while planning a recent infrastructure project, the utility considered a 25-year bond issue. The city's approval agency said, however, that a 20-year bond was preferred based on the project saving the city $1 million dollars on a $2.5 million job. Obtaining a different viewpoint in a project will almost always help.

LABOR AND INCENTIVES. Incentive plans can be effective cost-saving tools. The city of Indianapolis, for example, has an incentive plan for a solid waste division that could serve as a model (Carter, 1996). The Indianapolis program budget during the incentive plan was reduced from $43.9 million (in 1994) to $26.8 million (in 1996). In this particular case in 1993, city routes, private routes, and franchise routes were combined in 10 districts, and solid waste collection for each division was put out for bid. Three districts were awarded to the city's solid waste division as low bidder.

The solid waste division employees became more efficient and productive, and began doing more with less:

- Daily route crews decreased from 27 to 17 crews;
- Daily route size increased from 750 homes per day per crew to an average of 1 200 homes per day per crew;

- Absenteeism and worker compensation decreased; and
- Competition level increased.

As a result, the division began operating like a business. Employee salaries were frozen for the 3-year period. To offset the salary freeze, a one-time savings incentive check (for each of the 3 years) was given for each year proven savings occurred.

OUTSOURCING. Ultimately, the utility is accountable for three things: effectively providing services; efficiently allocating financial, staffing, and other resources needed for services; and achieving intended results (Carter, 1996). Economic and public pressures are forcing utilities to provide maximum service for the money invested, focus on customer satisfaction, and provide utility employees and managers with the tools they need to do their jobs. Several factors are driving competitive service delivery, including

- Desire to improve customer service,
- Cost containment,
- Focus on core services,
- Avoidance of major capital costs,
- Inability to hire and retain qualified staff,
- Flexibility to adapt to change, and
- Growing success and experience with outsourcing.

There are three basic steps for considering competitive services. These include identifying areas of competitive opportunities, evaluating options and preparing bid packages, and selecting and monitoring the contractor. When identifying bid opportunities, objectives and criteria must be established, data must be assembled and organized, roles for private and public sector must be defined, and opportunities must be evaluated. Evaluations should include a detailed cost analysis, including ABC analysis and contract manager cost. Performance standards should include quality of service and timeliness. Contract monitoring should include on-site reporting, operational reports, user surveys, complaint monitoring, and cost review.

EVALUATION OF EXISTING COLLECTION SYSTEM SPENDING. This section addresses strategies for determining where to spend resources. Specific examples are included as a guide to what other utilities are doing to prioritize spending and a consultant's method on how much is enough. Information on how to evaluate existing wastewater collection systems is available in the literature (U.S. EPA, 1991, and WEF, 1994).

The city of Indianapolis uses a bond strategy group, which includes city staff and consulting engineer advisors, to develop draft plans for funding levels and how the funds are split. In developing the draft, critical areas (in the sanitary realm) and major areas of work focus are considered (Jacob, 1997):

- Advanced wastewater treatment (AWT):
  - — Rehabilitation and repair,
  - — Expansion, and
  - — Modification.
- Combined sewer overflows and water quality:
  - — Structural,
  - — Rehabilitation and repair,
  - — Combined sewer overflow mitigation,
  - — Water quality improvements,
  - — Environmental remediation,
  - — Public health,
  - — Maintenance, and
  - — U.S. EPA requirements.
- Barrett law program (Indiana statute for funding Collection Sewers):
  - — Public health and
  - — Orderly expansion.
- Inflow/infiltration reduction program:
  - — Structural
  - — Rehabilitation and repair,
  - — Capacity recuperation,
  - — Public health,
  - — Overflow reduction, and
  - — U.S. EPA requirements (SSO policy is pending).
- Lift stations:
  - — Rehabilitation and repair,
  - — Expansion, and
  - — Modifications.
- Structural:
  - — Rehabilitation and repair and
  - — Public safety.

Figure 8.5 shows the breakdown (draft) spending in critical areas and the split. (Note that this breakdown for each city or utility will vary with the specific need of each. These needs will also vary with time.)

Over the past 4 years, the city has invested millions of dollars in these areas. The benefits of these investments to date have

- Decreased the city's future capital investment liabilities;
- Decreased (in many cases) the city's operating costs;
- Better used the existing systems, avoiding other capital costs;
- Improved the health and welfare of the public;
- Promoted orderly development; and
- Minimized a potential future rate (or tax) increase.

However, the more knowledge the city obtains about its collection system, the more aware it becomes about the magnitude of the needs in it.

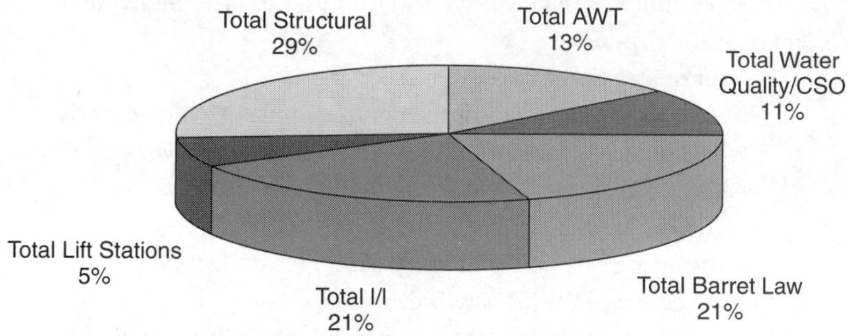

Figure 8.5 Breakdown (draft) spending in critical areas and split for Indianapolis, Indiana.

The following is a summary of the demonstrated capital needs in the city's system, along with the positive qualities (quantifiable and nonquantifiable) of why more investment today will actually further the achievements previously mentioned.

- Advanced wastewater treatment:
  - Operation savings are realized that allow continued reinvestment in the system;
  - More rehabilitation and repair options exist when they are implemented at owner's schedule;
  - Delays promote further deterioration, which in turn requires increased funding for rehabilitation and repair techniques;
  - Repairs dealt with at the time of failure significantly increase repair costs; and
  - The availability of capacity is a critical issue for continued growth.
- Combined sewer overflow and water quality:
  - Economic benefit through enhanced residential development opportunities with correlative tax base expansion is provided;
  - Community pride is promoted through highly visible projects;
  - Achievement of water quality standards and stream aesthetics is enhanced, thereby accelerating development of adjoining greenways for public use;
  - Compliance with federal and state regulatory requirements is promoted; and
  - Adverse health risks posed to public are reduced, thereby promoting recreational activities and economic development.
- Barrett law program:
  - Health risks are decreased because effluent runoff from septic systems is eliminated;
  - Revenue is increased because of new customers;
  - Installation of sanitary sewers can help promote more orderly development;

- Installation of sanitary sewers allows user to have more control over lot size;
- Water quality issues are addressed; and
- Program is also needed for the continued planned and orderly growth of the city and expansion of the sanitary sewer collection system.

- Inflow/infiltration reduction program:
  - More capacity can be reclaimed within the collection system;
  - The occurrence of SSOs will decrease;
  - As a secondary benefit, the line will be rehabilitated and repaired;
  - The occurrence of combined sewer overflows may decrease;
  - Increased health risks may be eliminated or reduced; and
  - Operational savings may be realized.
- Lift stations:
  - The occurrence of SSOs may decrease;
  - Increased health risks may be eliminated or reduced;
  - Operational savings that allow continued reinvestment in the system may be realized;
  - More rehabilitation and repair options exist when they are implemented at the owner's schedule;
  - Delays may promote further deterioration, which in turn may require increased funding for rehabilitation and repair techniques; and
  - Repairs are dealt with at the time of failure.

Recently, the Louisville–Jefferson County Metropolitan Sewer District (MSD) took steps to prioritize sanitary sewer systems tributary to various pumping stations and WWTPs that exhibited symptoms of excessive I/I (Garner, 1995). The process began through interviews conducted with the five divisions within MSD most affected by necessary responses to excessive wet weather flows in the sanitary sewer system. These groups included the Maintenance Division, Operations Division (Urban Areas Department), Physical Assets Division (Customer Service Department), Engineering Division, and Executive Office (Community Relations Administrator). The GIS group (or Louisville/Jefferson County Information Consortium [LOJIC]) was involved on a cursory basis because of its ability to retrieve and compile necessary geographic data. To facilitate the interviews, members of each group were asked approximately 25 questions regarding their perspectives on various issues, including responses implemented during wet weather; possible sources of I/I (such as roof leaders or leaky pipes); availability of data; and physical characteristics of the system, such as clay pipe versus polyvinyl chloride. After the interviews were completed, the information was recorded.

The next step involved an inventory of the separate sewer areas identified by the five divisions. A ranking of these systems was made by assigning weighing factors or pointing to symptoms such as customer complaints from basement backups, physical potential for environmental release, health and safety issues, and condition of receiving waters.

Water quality impacts were assessed through a general comparative analysis done by reviewing the receiving streams and comparing their general condition against others. Although the stream monitoring program is not geared to specific wet weather events, the "after effects" can often be seen by visual observations through water column analysis and by aquatic biological monitoring techniques. With water quality information in hand, a listing of sensitive or impaired stream reaches may be compared with potential SSO locations to help in the prioritization process.

After all weighing factors were assigned and totaled, additional data were added (such as the availability of television inspection data and flow monitoring). To finish the prioritization, the spreadsheet was distributed for comment or correction where necessary. It should be noted that priorities were established with potential for SSOs being only one of the six factors, although given a higher weighing than most factors.

The program is currently being implemented. An I/I abatement implementation plan is to be devised to ensure that the "worst first" are addressed based on a systemwide approach. Additionally, these prioritized areas can be integrated with other programs that are addressing known pollution sources (such as CSO, municipal separate storm sewer system [MS4] discharge, and nonpoint sources) to truly have a comprehensive watershed program that provides the community with the best results using available resources.

Establishing the cost to provide a given level of protection in a sanitary sewer system involves quantification of the following components:

- System rehabilitation to reduce wet weather flows;
- Relief sewer and pumping station construction to transport remaining peak wet weather flow after rehabilitation to prevent an SSO occurrence; and
- Treatment and storage of remaining wet weather flows.

System rehabilitation has been proven to be effective in reducing peak wet weather flow rates and total wet weather flow volume resulting from intense storm events. Cost-effective system rehabilitation, however, typically is not sufficient by itself without some combination of relief sewer and pumping station expansion or WWTP or storage facilities. All three of the cost components involve some level of ongoing maintenance; the success of many rehabilitation programs completed in the 1970s and 1980s has deteriorated because of the lack of an ongoing annual maintenance and rehabilitation program. This ongoing maintenance should be factored into the cost–benefit analysis by converting annual maintenance costs to total present worth.

It is often difficult to accurately quantify these costs under intense rainstorm conditions (greater than 1-year storm). It is typically far more difficult to quantify the benefits associated with providing a given level of storm protection. These benefits can include, but are not necessarily limited to, reduction in water quality violations in receiving streams and waters, reduction in damages to private property from basement backups, reduction in number of opportunities for primary human contact with SSOs, and reduction in number of occurrences with SSOs that result in closure of public recreation facilities or beaches and fishing waters.

Quantification of the effects of water quality violations often depends on computer modeling of receiving streams or waters. Computer modeling of water quality impacts of extreme rainstorm events is only as accurate as the model's ability to predict water quality conditions during and after events that happen infrequently. Thus, calibration of actual water quality sampling data with model results cannot rely on dry weather or even moderate stormwater quality sampling data; modeling of water quality conditions under extreme storm events must be calibrated with water quality sampling data gathered during those extreme events. Even some of the most sophisticated water quality models (including those developed under the old Clean Water Act Section 208 areawide water quality management plant programs) are lacking this type of extreme event calibration data.

Many communities do not have access to fully developed and maintained calibrated water quality models. In addition, the time and expense to develop and calibrate such models for extreme event conditions may not be justified given the uncertainty built into the modeling results under extreme event conditions.

Damages to private property from basement backups is also extremely difficult to project because of the number of variables involved, including the number of homes affected per SSO occurrence, the nature of the contents of each basement subject to backups, and the reduction in resale value of a home subject to basement backups. Quantifying damages for primary human contact with SSOs is also an inexact science, as is quantifying damages associated with closure of public recreation facilities and beaches and fishing waters.

One approach used during the last 15 years is based on traditional knee of the curve cost–benefit analysis, a curve used to determine level of service desired and costs to meet these levels (Hollenbeck, 1995). Instead of attempting to quantify benefits, an evaluation is made of where the breakpoint occurs in the marginal cost to prevent an SSO occurrence. An example of this analysis is shown in Figure 8.6. This analysis could be performed on a systemwide basis for small or medium-sized communities or on a separate basin or watershed scale for larger communities.

In every case where this methodology has been applied, some type of clear break point has emerged at which the marginal cost to prevent an SSO occurrence increased dramatically. The knee of the curve typically fell between a 1-year and a 10-year storm event. This would suggest a level of protection for an event that would occur on average between once per year and

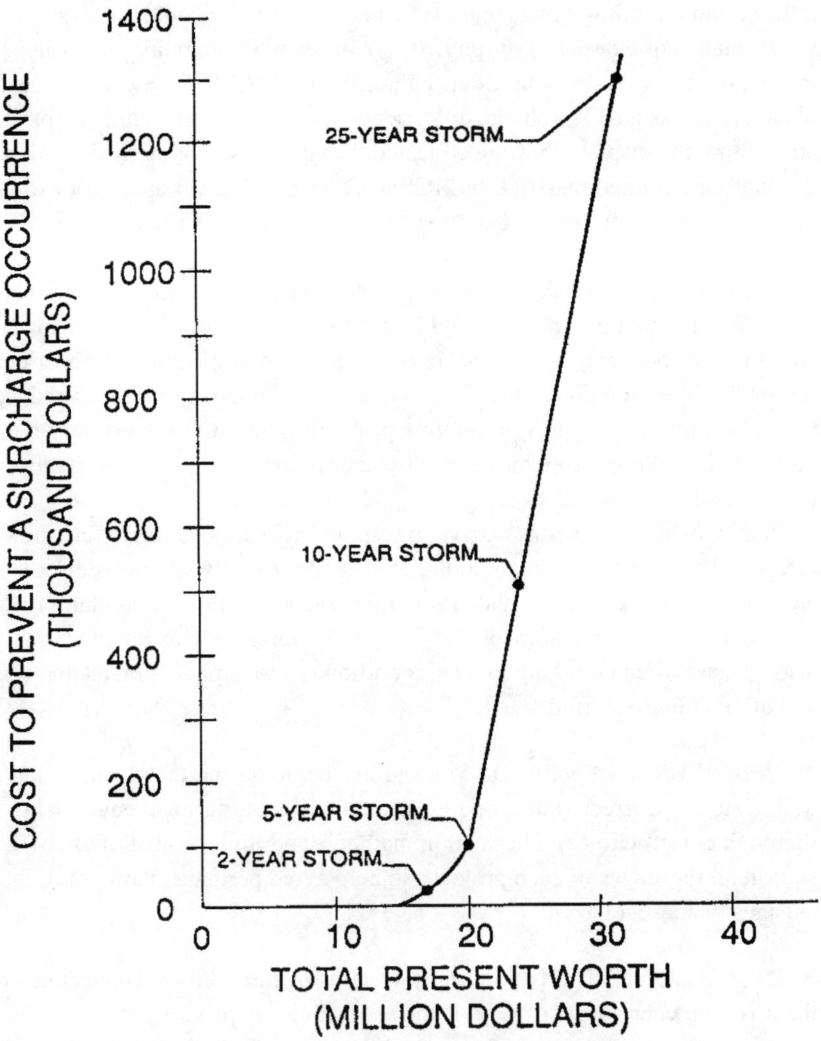

Figure 8.6 Relationship of cost for protection and design storm (Hollenbeck, 1995).

once every 10 years. As expected, this is a higher level of protection than is typically provided for CSO discharges.

CUSTOMER CONSIDERATIONS. The first rule of business (public or private) is to know the customer. For a utility, this includes everyone in the drainage basin. Whether they literally own the service lateral or pay a user fee, customers are bearing the cost of the wastewater treatment system. The key to a complaint-based system is to make sure the person complaining is truly calling for the correct reason. The best way for the utility to accomplish this is by educating everyone in the drainage basin about the system. Public education should occur at several points. Newspaper articles, public relations meetings, and direct mailings can give the users of the system small bits of information regarding things that they can do to help the collection system work better. For example (Murphy, 1995), at several community festivals throughout the year, Monroe County Pure Water (MCPW), New York, has a display, such as a large vehicle, to attract people's attention. While people are visiting the display, they are educated through personal interaction, visual displays, and printed material. In MCPW's combined sewer areas, for example, the need to keep leaves and other debris off of catch basin grates is stressed.

The second part of public education occurs when a customer calls for service. First, pointed questions should be asked to determine if the call actually warrants a response by a crew. Second, the caller should be told of his or her responsibility in relation to the sewer system and what can be done to reduce the current problem or prevent a future problem. The MCPW has reduced the number of field responses for service by approximately one-third through this method.

Finally, when a crew does arrive in response to a service call, a few minutes should be taken to explain to the homeowner exactly what was discovered or what service was provided. Figure 8.7 is a portion of a brochure that MCPW leaves with the homeowner for future reference. This brochure can also be used when crews are in a neighborhood or to explain a maintenance activity that has occurred.

NEW SEWERS. Effective O&M programs begin during the planning and design stages. Correct system sizing, selection of materials and equipment, thorough construction and inspection methods, and involvement of O&M staff in all the stages of each project can help avoid permanent system deficiencies that affect O&M.

SERVICE CONNECTIONS. Many systems include service connections to the private owner's right of way. If the sewer lines on private property were to be included in the municipal system, it would more than likely double or triple the amount of pipe maintained by the utility. It has been estimated that 10 to 40% or more of the I/I in a sewer line can be linked to laterals. This is a major problem to most systems because I/I uses up capacity and shortens (hydraulically) the life of the sewer.

Although economic analysis has shown that many private section I/I sources are cost effective to remove, there has been a reluctance to pursue this alternative largely because of legal and sometimes political difficulties associated with removal of I/I sources on private property. In the state of Ohio, a specific law was established to allow county commissioners to adopt rules requiring private property owners to disconnect stormwater inflows to sanitary sewers, disconnect nonstormwater inflows, and reconnect or relocate any such disconnected inflows. This law also allowed county commissioners to use sewer district funds, county general fund monies, and grants and loans for the cost of disconnection, reconnections, or relocations or for payments

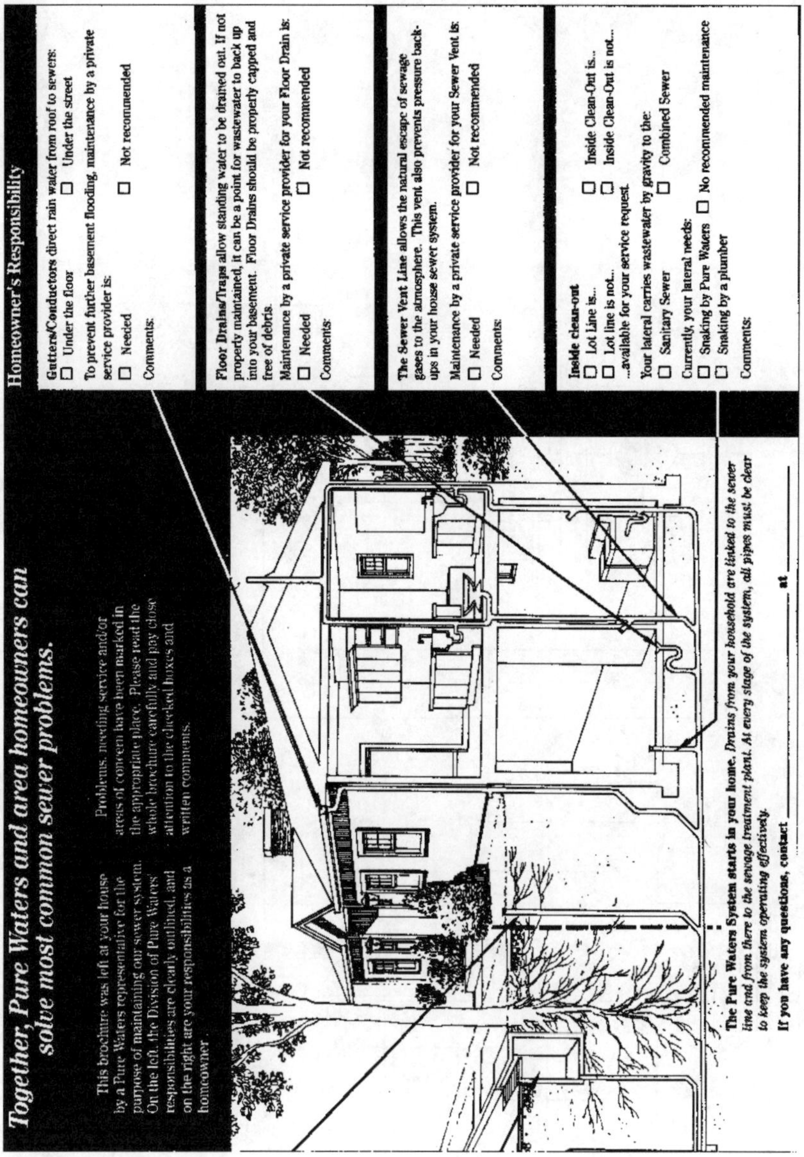

Figure 8.7 Sample literature (Murphy, 1995).

to property owners or contractors hired by property owners (competitive process where county crew specifies maximum amounts for each property).

Johnson County Unified Wastewater, Overland Park, Kansas, has developed a comprehensive private I/I removal program (Thoman, 1993). As shown in Figure 8.8, the program includes a disconnection process, a reimbursement process, an appeal process, and provision for obtaining access to the premises. Included in this program are county resolutions and a maximum reimbursement schedule for 12 items:

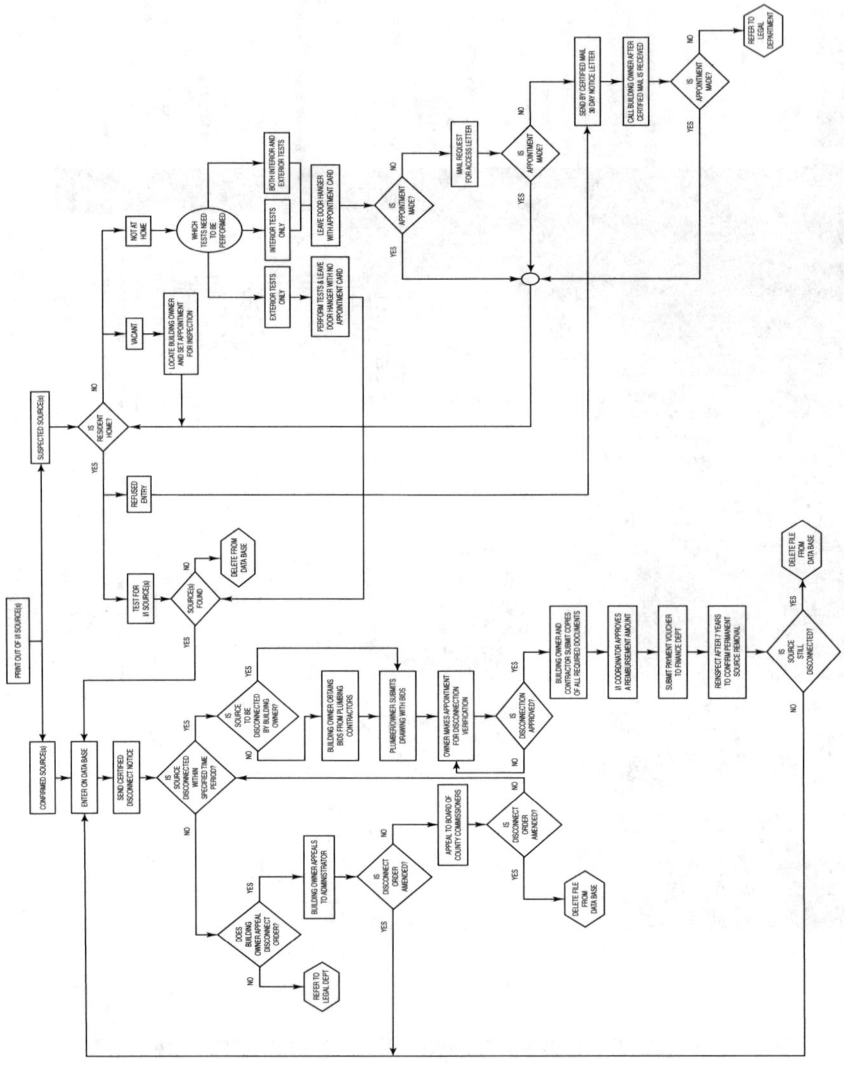

Figure 8.8 Johnson County, Kansas, unified wastewater private I/I flow chart (Thoman, 1993).

- Directly connected storm sump pump,
- Directly connected combination sump pump,
- Foundation drain sump pit,
- Combination foundation drain sump pit,
- Area drains (driveway, window well, patio, yard),
- Area drains (basement entry),
- Downspouts,
- Cleanouts,
- Auxiliary backup pump (separate authorization),
- Buried solid discharge pipe,
- Buried shallow lateral pipe, and
- Buried deep lateral pipe.

A step-by-step procedure for establishing a private I/I removal program is shown in Figure 8.9.

Lateral Repair Policy (Private Property). Shortly after a review of collected data, discussions were held with the utility board's staff to advise them of the significant effects that the lateral defects appeared to be having on this area and on other sectors of the collection system. This initiated a comprehensive reexamination of board lateral policy (U.S. EPA, 1995).

The board's lateral rehabilitation policy historically placed sole responsibility for lateral maintenance from the collector to the house with the private property owner. When lateral service problems were discovered, the property owner was notified and directed to make the correction. The board had made provision for repairs to the lower lateral if the lateral repair was associated with a capital improvements rehabilitation project. However, initial capital improvement projects failed to address the upper portion of lateral service defects.

After advising the board's staff of the urgent need to address the upper lateral problems if their goal of significant I/I reductions were to be realized, it was agreed that the board's policy should be revised. A more aggressive policy was recommended and approved. Private property owners would make I/I repairs detected on their laterals or risk having their water service terminated. The board gave the property owner the option of permitting the board's staff to make the repair to the lower lateral if the repair cost were reimbursed by the property owner. If the owner elected to use board restoration services, any other owner lateral defects discovered in the course of the work would also be repaired. A maximum cost ceiling to the property owner for lower lateral repairs was established at $1 200; the board assumed any costs more than this amount. The board also decided to offer property owners interest-free financing for owner-incurred lower lateral repair costs up to the maximum amount, with repayment to be made during a 4-year period. In addition, as further incentive to property owners to resolve lateral problems, the board would replace missing or broken cleanout caps for a cost of $13.52, which is far less than local area plumber fees for the same simple procedure.

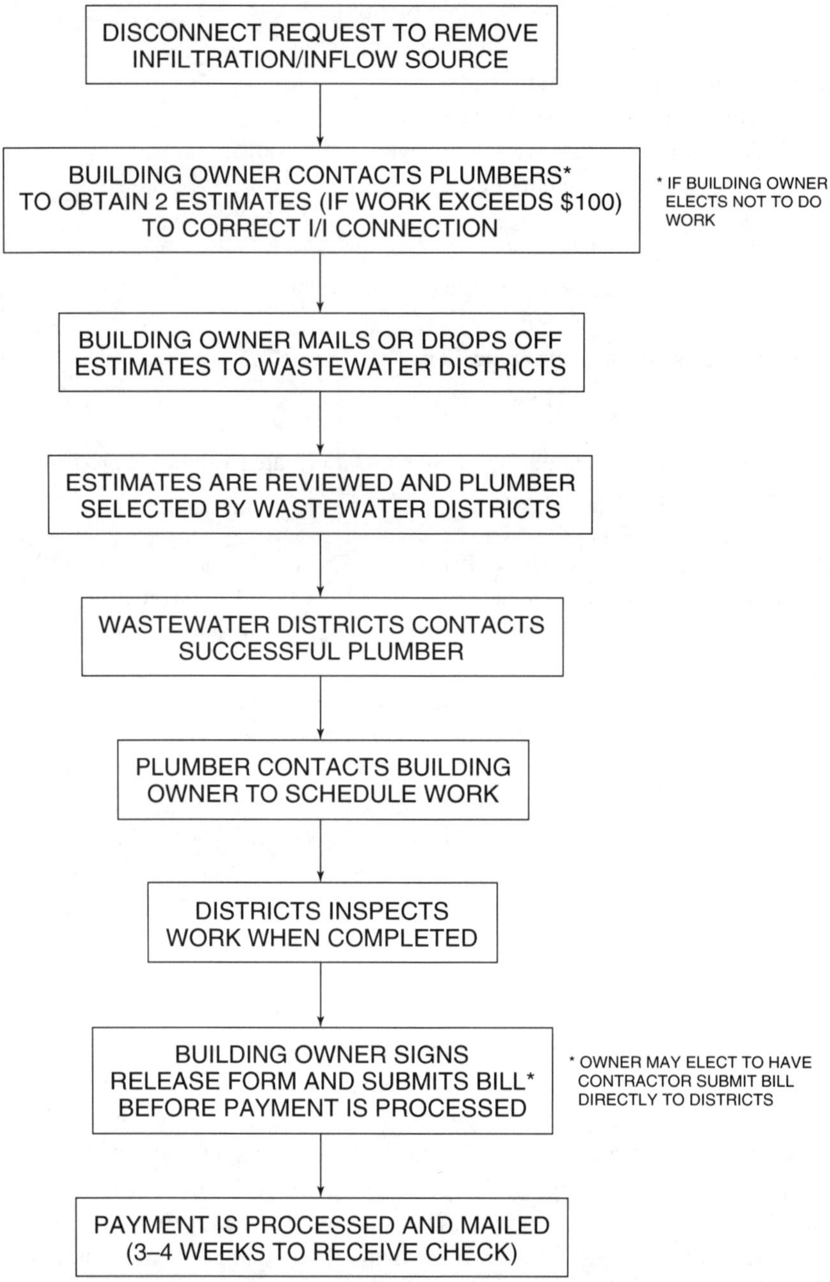

Figure 8.9 Step-by-step procedure to remove I/I source (U.S. EPA, 1995).

Under the current policy, a property owner retains the option of repairing any private defect. If the property owner chooses that option, a board inspector will go to the site and conduct a visual inspection of the work performed. If the repair is complete and of acceptable quality, a "thank you" letter is sent to the property owner. If the repair is not acceptable or another defect is located, a second notice is generated stating what the owner must do to bring the repair up to board standards.

When a property owner fails to respond to an initial 60-day notice of lateral repair requirements, a second 10-day notice is generated advising of possible interruption of service. If the property owner has not responded within the specified time frame, water service is terminated.

Basement Backup Abatement. In many cases, it takes years to determine and eliminate or reduce rain-induced I/I so that basement backup can be abated. In some locations, interim basement flooding programs have been developed until the I/I can be abated. These programs are sometimes tied to a lateral repair policy and storm disconnection program.

One program included requirement up to a maximum payment limit to remove downspouts and sump pumps (connected to stormwater source) and to repair deteriorated drains (service lateral to city sewer). Technical details included material and installation information, O&M program, inspection procedures, and plumber's guide and payment agreements. The program basically involved installation of shutoff valves and check valves between the house and sanitary sewer main. (Note that it is highly recommended that an engineer evaluate such installations for possible flotation of the house. The customer's failure to keep functional private protective devices may not excuse the city's [utility] liability for a sewer backup.)

MANHOLES. Manholes are often ignored or forgotten as I/I sources. Contribution of I/I varies widely among different sewerage systems (6 to 62%). Many of the problems are associated with the freeze-and-thaw zone, inlet pipes, and bench cone connection. Structural degradation will occur with three dimensional displacement. Manhole rehabilitation methods are described in the referenced literature (U.S. EPA, 1991, and WEF, 1994).

MAINLINE/PUMPING STATION PERFORMANCE. The conventional view of sewer performance has been that it is static and that it can be modeled and predicted with some accuracy (Petroff, 1995). Although this is true in some cases, it is not true in the majority of cases. Sewer system performance is far more dynamic than previously thought. During analysis of sewer flow over extended periods of time, performance of the sewer system constantly changes as blockages and constrictions appear and disappear, silt flows shift, new debris is thrown into the system, and bottlenecks are blown through the system.

The implications of this dynamic performance are profound for sewer modeling. Previously it was thought that if the design configuration of a sewer system was known, it could be modeled. Increasingly, this view is becoming obsolete. Every sewer has five different capacities:

- The capacity as it was designed,
- The capacity as it was actually constructed,
- The capacity that it has with present maintenance levels,
- The capacity because of downstream restrictions, and
- The capacity because of upstream I/I.

A sewer model can only describe the capacity as the sewer system was designed. Sewer modeling has changed so that both a "snapshot" of the sewer system and dynamic model of the system are available.

Sewer models were developed because accurate performance data on sewer systems were simply unavailable. Therefore, the only choice for engineers was to model what could not be measured. Now that what can be measured is more accurate than what can be modeled, the modeling will have to change to become more of a predictive tool.

REAL ESTATE ACQUISITION. Planning for the construction of sanitary sewer facilities requires that some special considerations be given during the design phase so that minimal problems will arise during acquisition of necessary land, whether as easements or in fee (land bought or owned outright). Improper planning can have serious effects on both project budgets and schedules (Benedict, 1997).

The choice of property to locate a WWTP on or align a sanitary sewer across is often a very different task from that which other utilities face, mainly because of the desire to construct sanitary sewers in a "gravity flow" mode whenever possible. Therefore, the selection of land is often limited by existing geographical and topographical features both occurring naturally and constructed. To limit the damage done during installation, it is desirable to avoid removing mature trees whenever possible. However, options are sometimes limited when a limited area is available to obtain gravity service. Therefore, trees and other plant material, flower beds, and shrubs, and sometimes physical structures such as garages, decks, and patios, will need to be removed. These considerations must be made when designing both the permanent and temporary construction easements. Both the immediate construction of the project and future maintenance and possible repair of the facilities must be considered.

Design of the sanitary sewer lines themselves and any needed easements should take into consideration both the cost of construction and cost of the land. Consideration must also be given to any obstacles that are in the path of construction and to determining the cost to remove them (including damages to the property owner), if they cannot be avoided. Both of these considerations are related to the formation of the project budget.

Properties will differ in value depending on many variables. The value of the land and therefore the value of taking the land is best determined by a certified appraiser. How a property is zoned can have a significant effect on the value of the taking. The variation in class (such as residential, commercial, and industrial) will often determine such things as setback limits or other building restrictions that may or may not affect the value of a partial taking, such as an easement. Specific deed, or other subdivision, restrictions may have the same effect. Depending on their location, species, and condition, the removal of trees or other vegetation can have a significant effect on valuation. Likewise, any structures such as garages, patios, or decks that must be removed will create additional expense for the project either in the form of payment for appraised value or cost to replace.

Therefore, a complete analysis of the project routing should be done to determine all alternatives, and a study should be done to determine the most cost-effective alignment to design. It must be remembered that the more disruptive and traumatic the construction is to property owners, the more likely they are to resist the taking. For private developments, this can mean lengthy delays, increased payments to property owners, or even abandoned projects, all of which have budgetary effects. For a public agency, with the right of eminent domain, the cost of litigation and the resulting slip in schedule can create a higher cost for acquiring the land.

The acquisition of property in fee for the location of either pumping stations or WWTPs requires additional considerations. Depending on the area needed, full or partial taking of the subject property could influence appraised values of the remaining property and therefore the amount due to the owner. In addition, it is always wise to conduct an environmental impact study before the execution of any sales agreement. Depending on the size and location of the property taken, it may be necessary to have an archeological survey or endangered species survey completed to reduce the possibility of dealing with costly delays or possible cancellations after the sale is complete. The most reliable means of accomplishing a review of all these concerns is to work in advance with an experienced land acquisition agent and a qualified appraiser. (For more information concerning the various aspects of professional land acquisition, contact the International Right of Way Association, 13650 S. Gramercy Place, Gardena, California 90249.)

PROACTIVE MAINTENANCE MANAGEMENT. One indication of how a system is operated and maintained is system performance (Arbour, 1997). Reactive maintenance—that is, waiting until a failure occurs before performing maintenance—will result in poor system performance, especially as the system ages. Proactive maintenance, where maintenance is performed on a periodic (preventive) or as needed (predictive) basis, will result in improved system performance, except when major chronic problems are the result of design and construction that cannot be rectified through O&M of the system.

To reduce O&M-related SSOs, changes must be made in the way collection systems are operated and maintained. To do that, managers need to know what factors affect collection equipment and systems and what the effect is on collection systems performance. The failure rate of a piece of

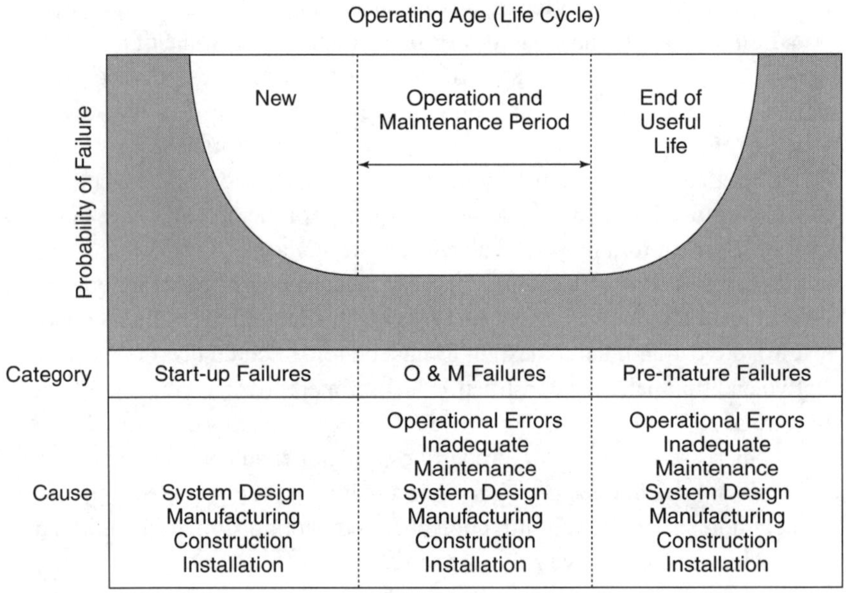

Figure 8.10 Equipment system failure rate as a function of age.

equipment (such as a pump) or a system (such as a gravity sewer) as a function of age is illustrated in Figure 8.10. The probability of failure is high during startup, is stable during the O&M period, and begins to increase as the useful life is reached.

Startup failures are caused by errors in design, manufacturing, construction, and installation. Unfortunately, many of these errors carry over into the O&M period and affect the performance of the system. A pump that is installed in a station with inadequate net positive suction head available is an example of a design problem that is nearly impossible to rectify after construction is complete. Hydrogen sulfide production and the resulting odor and corrosion problem would be another design problem. During the O&M period of the system, failures occur as a result of operational errors and inadequate maintenance of the system, in addition to the cumulative problems from the startup phase. Finally, increases in premature failures that are costly to repair begin to increase, signaling the end of the useful life of the equipment or system. The cumulative effects of failures are

- Substandard performance such as bypasses, overflows, reliability, and equipment availability;
- High O&M costs (reactive maintenance is always more costly than proactive maintenance);
- Reduction of the useful life of the system and equipment; and
- Higher rehabilitation/replacement costs.

An effective O&M program that examines the causes and effects of the system and equipment failures will extend the life of the collection system and reduce the number of O&M failures that result in SSOs. One approach to collection system O&M is asset management.

Asset Management. The collection system is a capital asset in which the community has invested. Figure 8.11 shows replacement costs of a collection system based on \$100/ft and \$200/ft (\$330/m and \$660/m). In a small system of 100 miles (160 km), the investment ranges from \$52.8 million to \$105.6 million. Managers should begin to operate and maintain the collection system as a business with a requirement for asset management. Instead of being treated as overhead, O&M should be considered a value-added activity, meaning it adds value to the original investment by maintaining capacity and extending its useful life.

A typical component of asset management is consideration of life-cycle cost, both as a design parameter and as a guide to O&M needs. The life-cycle cost of a piece of equipment or system is the entire cost during its lifetime. Life-cycle cost has two components: acquisition cost and ongoing cost. The acquisition cost is a one-time cost for the initial purchase. Ongoing costs are those costs that occur continuously for O&M, such as energy, repair, component replacement, and rehabilitation, from the time the system is started until the end of its useful life. The ratio of life-cycle costs to acquisition costs can vary widely depending on the equipment or system. The U.S. Navy for example, uses a ratio of 2.5 for its capital ships. A 1970 study of consumer products showed ratios from 1.9 to 4.8. Another way to consider this ratio is that the ongoing costs to operate and maintain are 1.9 to 4.8 times the origi-

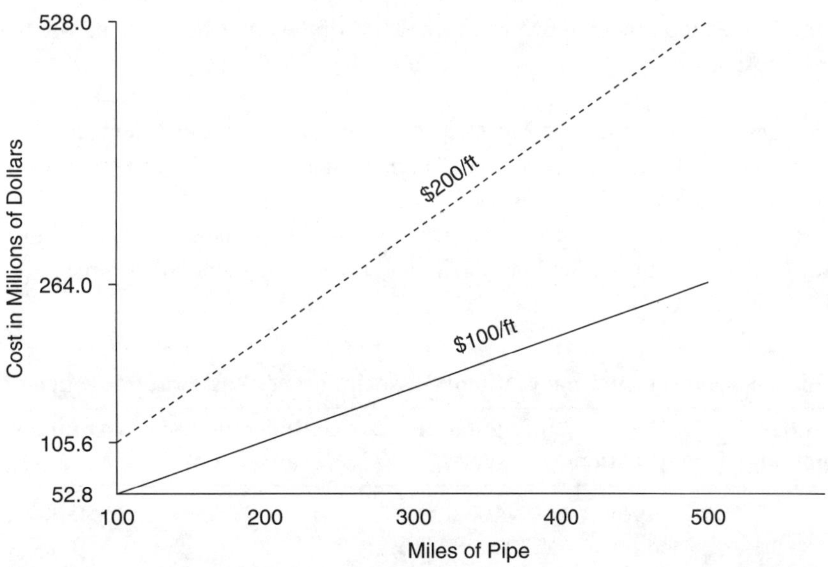

Figure 8.11 Collection system replacement cost.

nal acquisition cost. While this type of analysis is typical in private sector maintenance programs, it is not typical in public sector collection system O&M management. The O&M budgets for collection systems to provide the level of maintenance necessary to protect the capital investment and reduce SSOs that are O&M related are typically underfunded.

In 1994, U.S. EPA funded a statewide survey of the 61 member agencies of the Illinois Association of Wastewater Agencies. Twenty-four member agencies responded and were classified as small, medium, or large depending on the population served. Data from the survey are summarized in Table 8.6. Of these agencies, 24 reported having a preventive maintenance program in place, with 17 saying it was adequate and 7 reporting that it could be improved with additional resources. To determine spending levels, the annual maintenance budget was divided by the population served. Two agencies reported spending more than $25 per person served; 1 agency reported spending $15 per person served; 6 agencies reported spending between $5 and $10 per person served; and 15 agencies reported spending less than $5 per person served. While there is insufficient information provided in the report to draw any conclusions, there is a strong possibility that not enough money is being provided in a majority of these agencies to maintain the assets represented by their collection systems. This could even be the case of the two agencies at the high end, if the ratio of proactive maintenance costs to reactive maintenance costs is low.

A 1995 survey on water and wastewater user rates of the top 100 metropolitan statistical areas in the U.S. reports that in 1994, the average residential charge for wastewater was $16.24 per 7 000 gal ($0.61/m^3) (Ernst and Young, 1995). At the low end, the cost was approximately $3 per 7 000 gal ($0.11/m^3), and at the high end, just over $35 per 7 000 gal ($1.32/m^3). This survey provides an illustration of the wide range in user rates. When rates are maintained too low to allow adequate O&M of the collection system, SSOs will result.

Program Requirements. Most states have regulatory requirements for O&M of wastewater facilities. The language may be specific or very broad in scope, but typically includes collection systems as part of wastewater facilities. In addition, most agencies have sewer use ordinances governing the discharges into the collection system, including I/I from private sources,

Table 8.6 Survey summary, Illinois Association of Wastewater Agencies.

Number reporting	Classification	Population served	Number reporting SSO as a problem	Annual maintenance budget range
7	Large	>100 000	100%	$50 000–$1.5 million
10	Medium	40 000–99 999	80%	$50 000–$1.2 million
7	Small	<39 999	86%	$4 000–$580 000

which are known major contributors of I/I. Any analysis of federal regulations should include a state-by-state evaluation of existing regulations and whether they are being enforced.

The components of an effective collection system O&M program include

- Proper maintenance of collection systems and equipment,
- Correct operation of collection systems and equipment,
- A method to detect collection system defects and deterioration and a plan to repair them before failure,
- Identification and correction of design-related problems so that they are not carried over into new projects, and
- Development of employees to the level required to support the O&M program.

An effective O&M program requires that senior utility management, and the elected officials that have budget authority, understand the cost of the community's investment in the collection system and the need to protect that investment by providing adequate resources for O&M of the system. They must understand the consequence of O&M-related SSOs in terms of regulatory liability and the safety, health, and economic development of the community. Keeping user rates low by not maintaining the collection system is a short-term strategy for future disaster because the users will eventually have to pay for excessive rehabilitation or replacement costs as premature failures accelerate and the useful life of the system is shortened. In systems where chronic problems exist, additional resources may be required to implement an adequate O&M program because most of the existing resources are dedicated to reactive maintenance activities.

Those involved in collection system O&M must have a clear understanding of the concepts and principles of system and equipment O&M to implement an effective O&M program. They should know what is in the system, where it is, what condition it is in, and what must be done to maintain system functionality. System performance indicators should be developed to determine if the system is being operated and maintained in the most cost-effective manner. Throughout the organization, there must be a long-term commitment to shifting from a reactive maintenance philosophy to a proactive one. The O&M needs must be clearly documented and communicated to senior management and elected officials so that rational decisions can be made.

TRACKING PROGRESS

PERFORMANCE MEASURES. An asset management plan focuses on both the inputs to the process (investments) and the outputs (problem solved). Subsequently, both inputs and outputs must be monitored to assess performance.

As costs increase because of the need to comply with state and federal regulations, measurement of services, efforts, and accomplishments (SEA) becomes increasingly important to help ensure that resources are used economically, efficiently, and effectively (Government Accounting Standards Board, 1990). Wastewater authorities have recognized this need for better monitoring through SEA measures because of increasing public awareness of wastewater management issues, defense of ever-rising usage rates, and efforts to control maintenance costs. Some SEA indicators for wastewater treatment are given in Table 8.7.

When considering the reporting of SEA measures, it is important to examine the needs of information users. Groups that could use the data for evaluating SEA are the general public, industry, management, governmental legislative and oversight bodies, and the financial community.

The amount of information presented in cities' (utilities') annual reports on their wastewater activities typically ranges from nothing to several pages. Information provided includes a description of the department, its objectives, and a detailed listing of performance and workload measures. Generally, more outcome work indicators are presented than efficiency indicators.

Input indicators selected for wastewater include total cost of operations, cost per capita of wastewater treated, kilometres (miles) of infrastructure (pipeline), and number of employee hours. These indicators should provide information about the costs of the system and allow comparisons with similar-sized systems.

Several of the output indicators considered are the kilometres of sewer pipe maintained, repaired, and inspected; kilometres of new sewer constructed; and number of new sewers connected. These indicators would provide information about what is being done to meet current and future needs.

The condition and safety of the infrastructure and the success of the maintenance program should be measured with outcome indicators, such as the number of main stoppages per kilometre (100 miles) of sewer main, the number of complaints, the number of days effluent exceeded state and federal standards, and the I/I ratio. By definition, *infiltration* is groundwater that gets into the wastewater system through cracks and *inflow* is stormwater that gets into the system through manholes and cracks. Because it is expensive to perform wastewater treatment on stormwater and groundwater that does not need it, high levels of I/I indicate inefficiency. The I/I ratio provides a relationship between wet weather flow through the system and dry weather flow. Because of weather changes, four I/I ratios could be disclosed, one for each season. Also, trend data are important for proper analysis; an increasing ratio indicates that the sewer lines are deteriorating.

Another element of efficiency is customer service. An indicator that could point to a prompt response for customers' complaints is the percentage of repairs completed within a set period of time. Some utilities feel that performance measures can only be measured accurately if periodically checked by a third party outside the utility. This will vary with the utility.

Table 8.7 Recommended SEA indicators for wastewater treatment.

Indicator	Rationale for selecting indicator
Inputs:	
Total cost of operations Cost per capita of wastewater treated	To allow comparison of costs to other departments and other wastewater entities
Number and treatment capacity of WWTPs and level of treatment provided by each length of infrastructure (pipeline)	To provide a picture of the size of operations for which the entity is responsible
Number of employee hours	To indicate time spent on providing the service
Outputs:	
Length of sewer pipe maintained, repaired, and inspected (by geographic area) Percentage of length maintained requiring repair Percentage of above repaired this year	To indicate amount of infrastructure maintained, repaired, and inspected
Length of new sewer constructed Number of new services connected	To indicate the increase in the infrastructure to meet the needs of industry and the community in general
Number of service calls completed (by geographic area)	To indicate the level of work performed on existing system beyond general maintenance
Amount of wastewater treated (by treatment type) (BG): Primary treatment Secondary treatment Tertiary treatment	To indicate the level of work performed on existing system beyond general maintenance
Dry tons of sludge produced	To indicate the volume of dry sludge produced
Outcomes: Number of main stoppages per 100 miles of sewer main (by geographic area) Average service response time (in hours)	To determine how well the infrastructure is maintained
Number of complaints (by geographic area)	To indicate the quality of service, particularly from the customer's perspective

(Continued)

Table 8.7 Recommended SEA indicators for wastewater treatment (continued).

Indicator	Rationale for selecting indicator
Number of days effluent exceeded federal and/or state standards–number of violations of discharge permit (include reasons for noncompliance)	
Number of days influent exceeded WWTP capacity	
Volume effluent that did not meet federal standards/total volume processed through system	To indicate the ability of treatment process to remove pollution adequately
Quality of water in receiving body downstream from discharge	
Infiltration and inflow ratio	To indicate the condition of the infrastructure and the effectiveness of the maintenance program
Efficiency:	
Percentage of repairs completed within goal time	To indicate ability of the service group to clear calls within goal time
Wastewater treatment cost per volume treated (by treatment type):	
Primary treatment	To indicate the cost of providing the service and for comparison with other wastewater entities
Secondary treatment	
Tertiary treatment	
Solids disposal or use cost per dry kg	
Revenue from sales of by-products less costs	
Explanatory:	
Description of what the receiving body is used for	To provide information on the system's impact on the environment
Population served	To allow the reader to understand the size and demographics of the system
Area served	
Average daily flow/maximum daily treatment	To indicate the extent of excess capacity
Debt service coverage ratio	To show ability to pay debt
Projected needed capacity in 5 years/current capacity	To indicate the need for future expansion and funding
Total revenues from customer billings/total operating costs and debt service	To determine how much the city is subsidizing the department

EVALUATING THE EFFECTIVENESS OF WASTEWATER COLLECTION MAINTENANCE. Benchmarking. Benchmarking is defined as a method of comparing a utility or operation with a selected group of similar operations. Efforts have been made in the past to develop useful benchmarks. This section expands on those efforts in an attempt to provide benchmarks that can easily be used to compare and evaluate adequacy of personnel, review allocation of funds, and identify possible shortcomings.

Water Engineering/Management Survey. Thirteen major U.S. cities (with separate sanitary sewer systems) were surveyed to assess resources allocated to maintenance, environmental factors, the percentage of the system cleaned and video recorded, and the nature and number of SSO occurrences (Stamaker and Rigsy, 1997). (To ensure confidentiality, the utilities are identified by number only. Although utilities vary in size and operating conditions, these examples can be used for ideas on benchmarking or tracking in any system.)

Each city was asked to provide the length and age of the system, population served, number of service connections, budget for system maintenance, number of employees assigned to system maintenance, amount of system cleaned, amount of system video recorded, amount of system replaced or rehabilitated annually, number of stoppages, number of dry and wet weather overflows, number of backups into buildings, average dry weather flows, peak flows during wet weather, and annual rainfall amount. To provide a benchmark, the data gathered were then analyzed to compare the relative amount of resources dedicated to sewer system maintenance. The survey findings can be divided into three types: system characteristics, resources allocated to maintenance, and benchmarking ratios. Figure 8.12 shows the length of each of the 13 sanitary sewer systems that participated in the sur-

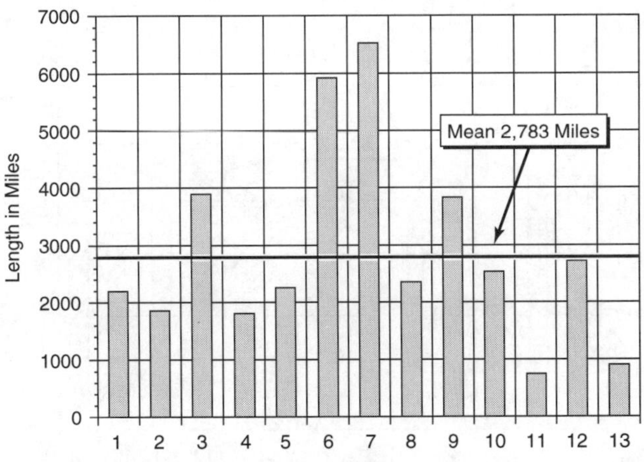

Figure 8.12 Sanitary sewer system length (Stamaker and Rigsy, 1997).

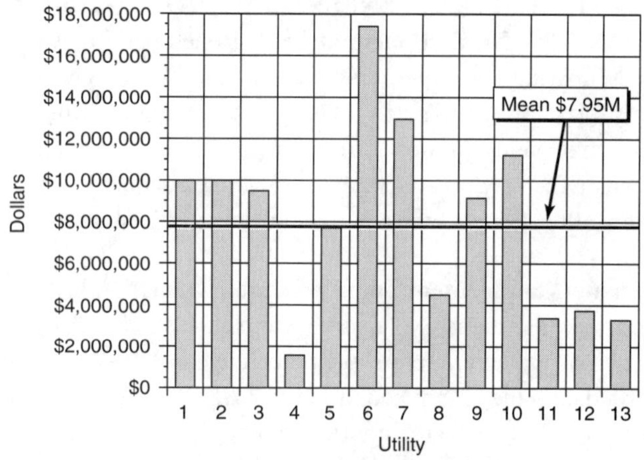

Figure 8.13 Annual maintenance budget (Stamaker and Rigsy, 1997).

vey. The average system length is 4 622 km (2 873 mi). Figure 8.13 illustrates the size of the annual sanitary sewer system maintenance budget. The average annual maintenance budget for these 13 utilities is $7 952 013. The ratio between the size of the annual maintenance budget and the length of the system is shown in Figure 8.14. The survey found that the average budget for these utilities is approximately $1 891/km ($3 042/mi) of sewer. Furthermore, the findings were relatively consistent, with most of the utilities not varying dramatically from the average.

Figure 8.15 illustrates related information comparing budget resources to system characteristics. It compares the size of the maintenance budget with the number of service connections to the system. The average utility included in the survey spends approximately $40 per service connection per year on maintenance. Figure 8.16 illustrates the relationship between the number of

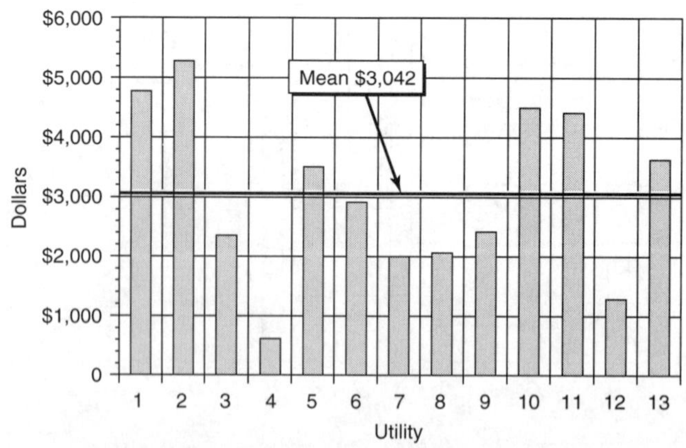

Figure 8.14 Maintenance budget per mile of sewer (Stamaker and Rigsy, 1997).

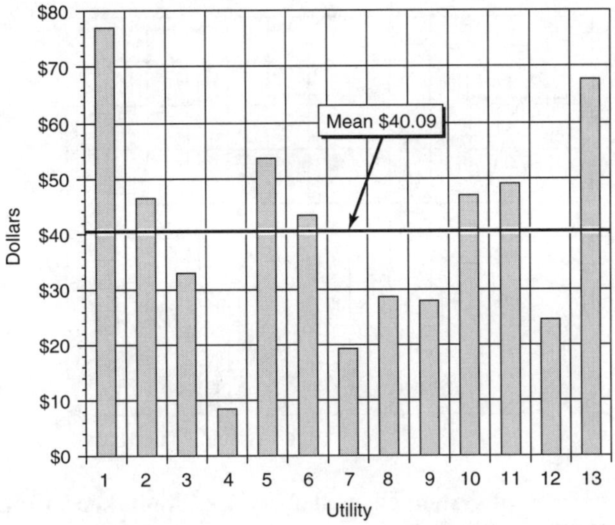

Figure 8.15 Maintenance budget dollars per service connection (Stamaker and Rigsy, 1997).

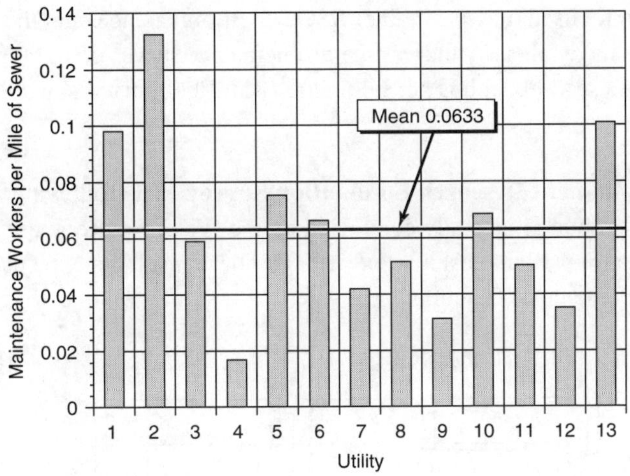

Figure 8.16 Maintenance workers per mile of sewer (mile $\times$ 1.609 = km) (Stamaker and Rigsy, 1997).

maintenance workers and the length of the sewer system. The average number of workers per mile is approximately 0.06, or 60 workers per thousand miles (0.04 per kilometre) of sewer system.

Figure 8.17 shows the percentage of the system cleaned annually by the surveyed utilities. The results indicate a great deal of variability, with an average amount of 24.02%. The percentage of the sewer system video recorded annually is also variable. The average percentage video recorded annually is approximately 4.8%, as seen in Figure 8.18.

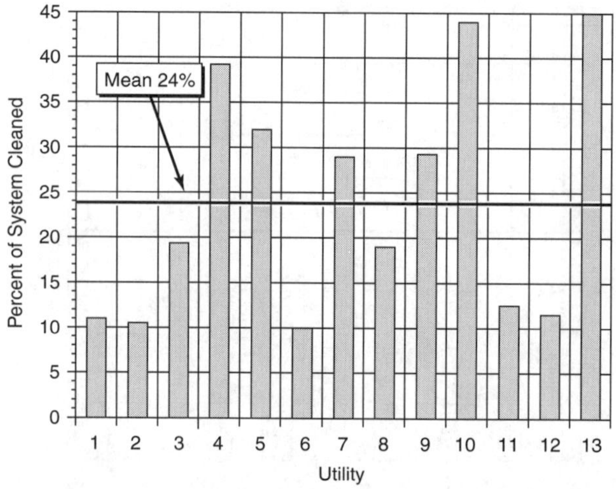

Figure 8.17 Percent of system cleaned annually (Stamaker and Rigsy, 1997).

Association of Metropolitan Sewerage Agencies Survey. In 1995, the Association of Metropolitan Sewerage Agencies conducted a survey of its members that focused on wet weather SSOs, their prevalence, and the steps these utilities have taken to address the problem. The survey found that an average of 513 sewer backups per kilometre (827/1 000 mi) of sewer line occur each year.

Water Environment Research Foundation Survey. The following is a list of benchmarks found in a study conducted by the Water Environment Research Foundation (Blankenship *et al.,* 1997). These results may be useful in evaluating maintenance function:

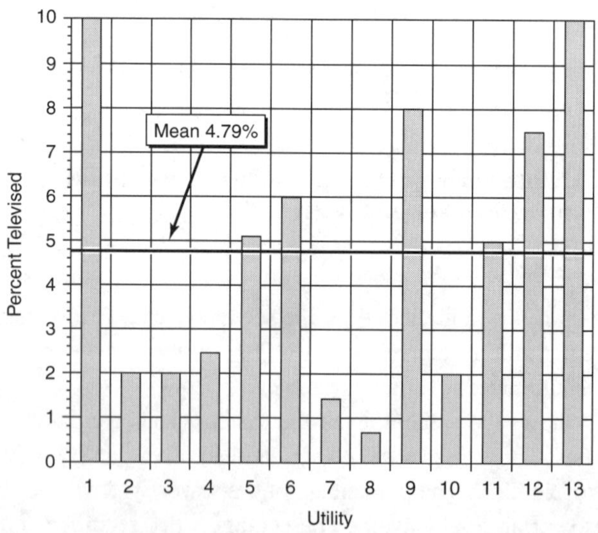

Figure 8.18 Percent of system video recorded annually (Stamaker and Rigsy, 1997).

- Overtime rate of less than 6 to 8% of total labor;
- Actual equipment available 95%;
- Actual equipment utilization greater than 85%;
- Maintenance workforce hands-on greater than 50%;
- Preventive maintenance greater than 50% of total maintenance budget;
- Preventive maintenance inspections completed greater than 99%;
- Work orders scheduled versus executed greater than 85%;
- Work orders unscheduled 15%;
- Maintenance cost decreasing 2 to 3% annually;
- Warehouse items outages less than 5%; and
- Warehouse requests completely filled when stocked greater than 99%.

REFERENCES

Allio, R.P., *et al.* (1994) Benchmarking: A Management Tool for Performance Improvement. *Water/Eng. Manage.*

American Water Works Association (1990) *Revenue Requirements.* Manual M35, Denver, Colo.

American Water Works Association (1991) *Water Rates.* 4th Ed., Manual M-1, Denver, Colo.

Arbour, R. (1997) The Importance of Your O&M Program in SSO Management. Collection System Specialty Seminar.

ASTM (1996) *Annual Book of ASTM Standards.* West Conshohocken, Pa.

Benedict, D.H. (1997) Land Acquisition Manager, Louisville & Jefferson County Metropolitan Sewer District.

Blankenship, L.; Liner, B.; Olstein, M.; Stevens, B.; Koch, C.; and Kochaba, T. (1997) Benchmarking Best Practices for Collection System Operation and Maintenance. Paper presented at Collection Systems O&M: Solving Today's Problems and Meeting Tomorrow's Needs, Water Environ. Fed. Specialty Conf., Kansas City, Mo., 5.

Carter, M. (1996) Indianapolis Department of Public Works, Training Session.

Ernst and Young (1995) *1995 Water and Wastewater Rate Survey.* Washington, D.C.

Finkle, R.W. (1995) Providing Better Service to Customers: The Role of Process Re-Engineering and AM/FM at the City of Wichita Water and Sewer Department. Paper presented at Sewers of the Future, Water Environ. Fed. Specialty Conf., Houston, Tex.

Garner, G. (1995) Prioritize Sanitary Sewer Overflows Based on System Impacts and Water Quality. Paper presented at Natl. Conf. Sanit. Sewer Overflows. Washington, D.C.

Government Accounting Standards Board (1990) Its Time Has Come. In *Research Report Service Efforts and Accomplishments.* Norwalk, Conn.

Hollenbeck, A.J. (1995) Seperate Sewer Overflows—Design Storm Protection Level. Paper presented at Sewers of the Future, Water Environ. Fed. Specialty Conf., Houston, Tex., 3.
Jacob, M. (1997) Department of Capital Asset Management, City of Indianapolis, Indiana, Personal Communication.
Kingdon, W.D. (1995) Water/Engineering and Management. Water Research Center.
Lior, S.K. (1996) Integrasting Data for Automation. *Water Environ. Technol.*, **8**, 7, 65.
Murphy, S. (1995) Complaint Based Sewer Maintenance. *Proc. Water Environ. Fed. Specialty Conf. Sewers of the Future,* Houston, Tex., 8-11.
National Association of Sewer Service Companies (1995) *Manual of Practices, Wastewater Collection Systems.* 1st Ed., Maitland, Fla.
OhioScoop (1997) **14,** 1.
Petroff, R.G. (1995) A Future Perspective. Paper presented at Sewers of the Future, Water Environ. Fed. Specialty Conf., Houston, Tex.
Sawatzki, J. (1996) *World Water Environ. Eng.*
Stamaker, R., and Rigsy, M. (1997) *Water/Eng. Manage.*, January.
Taylor, A. (1995) Strategic Fleet Management. *Public Works*, May.
Thoman, R. (1993) Johnson County Wastewater Private Infiltration and Inflow (I/I) Removal Program—Procedures Manual. Overland Park, Kans.
U.S. Environmental Protection Agency (1991) *Sewer Infrastructure Analysis and Rehabilitation.* EPA-625/6-91-030, Office Res. Dev., Cincinnati, Ohio.
U.S. Environmental Protection Agency (1995) Sanitary Sewer Overflow Workshop. Washington, D.C.
University of North Carolina (1987) Financing Water & Sewer Extensions in Urban Growth Areas, Current Practices and Policy Alternatives. Capital Program Planning.
Ward, J. (1994) *Am. City.*
Water Environment Research Foundation (1997) Development of Predictive Tools for Measuring Sewer Degradation. Contract No. 97-CTS-7, Alexandria, Va.
Water Environment Federation (1994) *Existing Sewer Evaluation and Rehabilitation.* Manual of Practice No. FD-6, Alexandria, Va.
Water Pollution Control Federation (1982) *Gravity Sewer Design and Construction.* Manual of Practice No. FD-5, Washington, D.C.; Manuals and Reports on Engineering Practice No. 60, Am. Soc. Civ. Eng., New York, N.Y.
Water Pollution Control Federation (1984) Financing and Charges for Wastewater Systems. 2nd Ed., Washington, D.C.
Woolpert Consultants (1997) Sanitation District No. 1, 28 Oct 97. Public Hearing, Preparing for 21st Century—20 Year Sewer Plan, Ft. Wright, Ky.

Appendix Defect Classifications Guide

This guide has been developed to assist users in the inspection and maintenance of wastewater collection systems. A detailed list of pipeline and surface defects or irregularities is presented with a coding system to provide a uniform means of identifying and recording system characteristics during pipeline inspection. In addition, a coding system for actions or responses to system deficiencies is included. Uniformly coding and recording wastewater system characteristics and responses to failures for the collection systems industry increases the efficiency of pipeline inspection and preventive maintenance.

The inspector can work in the field with the abbreviations list as a reference and quickly and efficiently record the pipeline characteristics by code and location on the inspection log. Maintenance personnel can then interpret pipeline characteristics by reviewing the log and a closed circuit television (CCTV) tape or photographs and using this guide. The maintenance department can then assign action/response codes to the system deficiencies that require correction.

The coding systems presented include typical characteristics and defects of wastewater collection systems and common responses to service problems. Therefore, it may be necessary to add specific items to the defect and action/response listings to completely evaluate the system being inspected.

ABBREVIATIONS OF DEFECTS

Code	Description	Unit
BBU	Basement backed up	ea
BDM	Brick demortared	m^2 (sq ft)
BEG	Begin inspection	ea
BMS	Brick missing	m^2 (sq ft)
CAB	Camera blocked	ea
CAS	Camera submerged	ea
CAW	Contractor at work	ea
CCB	Catch basin (storm drain) culvert blocked	ea
CFD	Catch basin full/debris	ea
CGB	Catch basin grate blocked	ea
CGG	Catch basin grate gone	ea
CIN	Contract inspection	ea
CKC	Cracking, circumferential	m (ft)
CKL	Cracking, longitudinal	m (ft)
CNA	Crew needs assistance	ea
COR	Corrosion, hydrogen sulfide	m (ft)
CRS	Crushed	m (ft)
DAC	Debris accumulated	m (ft)
DEP	Depression/void	m^2 (sq ft)
DEW	Dewater required	L (gal)
DFR	Dust for roaches	m (ft)
DIS	Distorted/out of round	m (ft)
DMG	Damage, other	ea
END	End of inspection	m (ft)
FLD	Flow depth	cm (in.)
GRS	Grease accumulated	m (ft)
IFL	Infiltration rate	L/s (gpm)
IFO	Inflow rate	L/s (gpm)
INS	Inspection needed	ea
INV	Invert damage	m (ft)
ISB	Internal sump blocked	ea
ITD	Inflow, tidewater	L/s (gpm)
JGB	Joint gasket bad	ea
JOP	Joint separated/open	cm (in.)
JOS	Joint offset	cm (in.)
LBK	Line blocked	ea
LDA	Line deflect above	m (ft)
LDB	Line deflect below	m (ft)
LDL	Line deflect on left	m (ft)
LDR	Line deflect on right	m (ft)
LOS	Lost article	ea

LSG	Line sagged	m (ft)
LSR	Line surcharge	ea
LTC	Line-type change	ea
MAJ	Minerals at joint	ea
MHA	Manhole above grade	ea
MHB	Manhole below grade	ea
MHC	Manhole covered/buried	ea
MHN	Manhole noisy/rattling	ea
MHO	Manhole cover open/gone	ea
MHP	Manhole to be paved	ea
MHS	Manhole surcharge/overflow	ea
MOW	Minerals on wall	m (ft)
MRC	Mechanism rusted closed	ea
MRO	Mechanism rusted open	ea
ODR	Odor complaint	ea
ODW	Overflow, dry weather	ea
ORP	Owner responsible	ea
OTP	Other problem(s)	ea
OWW	Overflow, wet weather	ea
PSP	Pump/pump station problem	ea
RAT	Rat infestation	ea
ROA	Roach infestation	ea
ROO	Root intrusion	m (ft)
SDM	Structure damage	ea
SER	Serviceable	ea
SLB	Service lateral blocked	ea
SLC	Service lateral crushed	m (ft)
SPL	Spalling	m (ft)
SPM	Part(s) missing	ea
SSR	Scheduled service	ea
TA1	Tap abandoned 12–3	ea
TA2	Tap abandoned 3–6	ea
TA3	Tap abandoned 6–9	ea
TA4	Tap abandoned 9–12	ea
TC1	Tap crushed 12–3	ea
TC2	Tap crushed 3–6	ea
TC3	Tap crushed 6–9	ea
TC4	Tap crushed 9–12	ea
TI1	Tap infiltration 12–3	ea
TI2	Tap infiltration 3–6	ea
TI3	Tap infiltration 6–9	ea
TI4	Tap infiltration 9–12	ea
TN1	Tap intruding 12–3	ea
TN2	Tap intruding 3–6	ea
TN3	Tap intruding 6–9	ea

TN4	Tap intruding 9–12	ea
TP1	Tap normal 12–3	ea
TP2	Tap normal 3–6	ea
TP3	Tap normal 6–9	ea
TP4	Tap normal 9–12	ea
TR1	Tap with roots12–3	ea
TR2	Tap with roots 3–6	ea
TR3	Tap with roots 6–9	ea
TR4	Tap with roots 9–12	ea
TRP	Trench to be paved	m^2 (sq ft)
TS1	Tap separated 12–3	ea
TS2	Tap separated 3–6	ea
TS3	Tap separated 6–9	ea
TS4	Tap separated 9–12	ea
TU1	Tap unknown 12–3	ea
TU2	Tap unknown 3–6	ea
TU3	Tap unknown 6–9	ea
TU4	Tap unknown 9–12	ea
UNK	Unknown situation	ea
VOL	Void outside line	m^3 (cu yd)
VRM	Vermin, general	ea
WMB	Water main broken	ea

Glossary of Defects

BBU—Basement backed up (ea) Wastewater from a sewer pipe is entering and flooding the basement or lower level of a building.

BDM—Brick demortared (m^2 or sq ft) In a brick sewer, the cement or mortar between the bricks is deteriorating, causing spaces between the bricks. This item is measured in square metres or square feet of brick surface demortared.

BEG—Begin inspection (ea) The point where the sewer inspection is started. Normally, this is the manhole where the closed circuit television camera is pulled into the system or, in man-entry sewers, where the inspector enters the system.

BMS—Brick missing (m^2 or sq ft) One or more bricks missing in a brick sewer, often at the invert (bottom) or crown (top), evidenced by a hole in the pipe wall. Sewers often have several layers or courses of bricks, so even though one course is missing, bricks still may be visible underneath. This item is measured in square metres or square feet of surface area with missing bricks.

CAB—Camera blocked (ea) The advancement of the television camera is stopped by a physical obstruction.

CAS—Camera submerged (ea) High flows in the pipe are immersing the closed circuit television camera lens and blocking camera vision. If this condition persists over a considerable length of the pipe, the inspection should be abandoned and resumed in a low-flow period or after the pipe has been dewatered.

CAW—Contractor at work (ea) A contractor is engaged in construction activities in or around the pipe, including work on the ground surface.

CCB—Catch basin (storm drain) culvert blocked (ea) The culvert or pipe exiting a catch basin (storm drain) is blocked, inhibiting flow through the pipe. This is evidenced by standing water in the catch basin.

CFD—Catch basin (storm drain) full/debris (ea) Trash has collected inside a catch basin, inhibiting flow out of the structure.

CGB—Catch basin (storm drain) grate blocked (ea) Trash has collected on the grate of a catch basin (storm drain), inhibiting stormwater entry.

CGG—Control basin (storm drain) grate gone (ea) The catch basin (storm drain) grate is missing, leaving a dangerous hole for passersby and vehicles and an entryway for debris that could block the system.

CIN—Contract inspection (ea) Inspection of piping system repairs done by an outside contractor.

CKC—Cracking, circumferential (m or ft) A break or crack in the pipe wall running laterally or from side to side along the pipe cross section. This defect is measured as the length of the crack in metres or feet.

CKL—Cracking, longitudinal (m or ft) A break or crack in the pipe wall running along the length of the pipe. It is measured as the length of the crack in metres or feet.

CNA—Crew needs assistance (ea) The crew requires additional manpower or special services because of a difficult inspection, equipment problems, emergency needs, or safety purposes.

COR—Corrosion, hydrogen sulfide (m or ft) Portions of the internal pipe wall have deteriorated as a result of hydrogen sulfide corrosion. This normally occurs at the crown (top) of concrete pipes and is evidenced by deteriorated sections of the pipe wall often resulting in the exposure of the reinforcing steel. This defect is measured in linear metres or linear feet of pipe affected by corrosion.

CRS—Crushed (m or ft) Sections of the pipe have collapsed due to structural failure, blocking flow and often stopping advancement of the television camera. This item is measured in linear metres or linear feet of affected pipe.

DAC—Debris accumulated (m or ft) Trash has collected in the pipe, inhibiting flow. This typically occurs at the pipe invert (bottom) and joints. This item is measured in linear metres or linear feet of affected pipe.

DEP—Depression/void (m^2 or sq ft) A hole or sunken spot on the ground surface in the vicinity of the sewer system. The cause of this depression or void formation is located and corrected to avoid future occurrences. This item is measured in square metres or square feet of surface area of the depression/void.

DEW—Dewater required (L or gal) If the flow in the pipe is so great that it is blocking visual inspection of the system, bypass pumping may be required. This reroutes the flow from an upstream manhole to some point downstream of the section being inspected, thus eliminating flow through the pipe. Dewatering is measured in litres or gallons of flow to be bypassed.

DFR—Dust for roaches (m or ft) If roach infestation is excessive, dusting of an insecticide may be required to clear the pipe. This item is measured in linear metres or linear feet of pipe needing treatment.

DIS—Distorted/out of round (m or ft) The cross section of the pipe has been deformed under external loads. Typically, a circular pipe has deflected to a horizontal oval or, in some cases, vertical oval. Cracking of the pipe often accompanies this condition. This item is measured in linear metres or linear feet of pipe affected.

DMG—Damage, other (ea) This item includes structural damage to the pipe that is not included in any other defect classification listed.

END—End of inspection (m or ft) The point at which an inspection of a pipe length is completed. This normally occurs at the manhole where the television camera is removed from the system or in manual inspections, where the inspector exits the pipe. This item is measured in total pipe length (in metres or feet) inspected.

FLD—Flow depth (cm or in.) A gauge of the amount of flow in the pipe being inspected, measured as the height of the water surface above the invert (bottom) of the pipe in centimetres (inches).

GRS—Grease accumulated (m or ft) A thick, oily substance is coating the pipe wall. This normally occurs at the average water level in the pipe. This item is measured as length (in metres or feet) of pipe affected.

IFL—Infiltration rate (L/s or gpm) Infiltration is groundwater entering the pipe from the exterior. Infiltration occurs at joints and cracks in the piping system and manholes. This item is measured in litres per second (gallons per minute) of water entering the system.

IFO—Inflow rate (L/s or gpm) Inflow is storm flow from the ground surface coming into the pipe from the exterior. The water enters the system through manhole covers, cracks in the curb or pavement, ditch flow, or other sources. This item is measured in litres per second (gallons per minute) of water entering the system.

INS—Inspection needed (ea) Inspection of a length of pipe is necessary because of suspected problems or routine maintenance.

INV—Invert damage (m or ft) Deterioration of the pipe at the invert (bottom) often characterized by cracking, voids, corrosion, or reverse curvature. This item is measured in linear metres or linear feet of pipe affected.

ISB—Internal pump blocked (ea) A pump in the system is not operating, causing the flow to back up into upstream pipes. This problem requires immediate attention by maintenance crews.

ITD—Inflow, tidewater (L/s or gpm) Tidewater inflow occurs in sewer outfalls where lake or ocean water is washing back into the pipe. This item is measured in litres per second (gallons per minute) of water entering the system.

JGB—Joint gasket bad (ea) The rubber material that provides a watertight joint seal is deteriorating. This is evidenced by leakage and pipe deterioration at the joints.

JOP—Joint separated/open (cm or in.) Two adjacent pipe sections have moved differentially in the vertical plane, causing a gap or space at the joint. This item is measured in centimetres (inches) of the opening at the joint.

JOS—Joint offset (cm or in.) Two adjacent pipe sections have moved differentially in the horizontal and/or vertical direction and caused shifting at the joint. This is evidenced by a discontinuity in the smooth interior surface of the pipe where there is a step or offset at the joint. Debris often collects in this area. This item is measured in centimetres (inches) of the offset or step at the joint.

LBK—Line blocked (ea) A stoppage in the pipe is inhibiting flow and typically preventing further advancement of the television camera.

LDA, B, L, R—Line deflect above, below, on left, on right (m or ft) The pipe wall has moved inward as a result of external loadings, often flattening the wall or causing a reverse curvature. Cracking often accompanies this condition. This item is measured in linear metres or linear feet of pipe affected.

LOS—Lost article (ea) A public report that a personal object has been dropped into the sewer system.

LSG—Line sagged (m or ft) A section of pipe has dropped or displaced vertically downward, causing a low spot in the pipeline. In other words, the slope of the pipe is not continually downhill and flow is collecting in a sagged area. This item is measured in linear metres or linear feet of pipe sagged.

LSR—Line surcharge (ea) Flow in the pipe is over capacity so that it backs up into house connections (service laterals) and upstream manholes. This condition can be caused, for example, by line blockages, excessive infiltration and inflow, undersized pipe, or downstream pump failure.

LTC—Line-type change (ea) The cross section of the pipe changes to a new size or a different shape. The type of change is indicated.

MAJ—Minerals at joint (ea) Mineral deposits have collected and caused encrustation at the joint. These minerals can form as a result of wastewater or soil properties. This item is measured per joint affected.

MHA—Manhole above grade (ea) The manhole cover is higher than the surrounding ground.

MHB—Manhole below grade (ea) The manhole cover is lower than the surrounding ground, allowing inflow to the sewer system.

MHC—Manhole covered/buried (ea) The manhole cover is not visible. Manhole covers are sometimes paved over or buried with earth so that they are difficult to locate.

MHN—Manhole noisy/rattling (ea) The manhole cover is not seated tightly and moves under traffic and rattles.

MHO—Manhole cover open/gone (ea) The manhole cover is not in place or is missing, leaving a dangerous hole for passersby and vehicles.

MHP—Manhole to be paved (ea) A manhole is to be abandoned and paved over to avoid a potential hazard and reduce inflow.

MHS—Manhole surcharge/overflow (ea) Flow in the pipe is above capacity, causing backup into upstream manholes. Backup is evidenced by a rising water level in the manhole submerging the outgoing pipe. This condition can be caused, for example, by line blockages, excessive infiltration and inflow, undersized pipe, and downstream pump failure.

MOW—Minerals on wall (m or ft) Mineral deposits have collected on the pipe wall and caused an encrusted area. These minerals can form as a result of wastewater or soil properties. This item is measured as linear metres or linear feet of pipe affected.

MRC—Mechanism rusted closed (ea) A tide gate, regulator, or other form of flow-control mechanism has rusted closed so that it cannot be removed or operated properly.

MRO—Mechanism rusted open (ea) A tide gate, regulator, or other form of flow-control mechanism has rusted open so that it cannot be operated properly.

ODR—Odor complaint (ea) Public complaints of unpleasant odors may indicate problems in the sewer system.

ODW—Overflow, dry weather (ea) During dry weather conditions, flow in the pipe is over capacity and backs up and surfaces in homes and businesses through house laterals or on the ground surface through manholes or leaks in the pipe.

ORP—Owner responsible (ea) Maintenance is required to correct a sewer problem that is the responsibility of the property owner. These items are normally related to the service lateral.

OTP—Other problem(s) (ea) This item covers problems that do not fall under any other defect classification listed. The inspection report describes the item.

OWW—Overflow, wet weather (ea) During wet weather conditions, flow in the pipe is over capacity and backs up and surfaces in homes and businesses through house laterals or on the ground surface through manholes or leaks in the pipe.

PSP—Pump/pump station problem (ea) Maintenance is required for pumps or pumping stations.

RAT—Rat infestation (ea) Rats were seen in the collection system (per sighting).

ROA—Roach infestation (ea) Roaches were seen in the collection system (per sighting).

ROO—Root intrusion (m or ft) Tree roots have entered the pipe. This condition occurs at joints and cracks in the pipe wall. Root intrusion can inhibit flow and promote further deterioration of the pipe. This item is measured in linear metres or linear feet of affected pipe.

SDM—Structure damage (ea) Cracking and deterioration of structures such as manholes, catch basins, storm drains, or valve pits.

SER—Serviceable (ea) No defects are visible during the inspection.

SLB—Service lateral blocked (ea) Stoppage in a house connection (service lateral) is inhibiting flow and causing wastewater backup.

SLC—Service lateral crushed (m or ft) A length (in linear metres or linear feet) of the house connection (service lateral) has collapsed. Flow is inhibited and surface disturbances are possible.

SPL—Spalling (m or ft) The interior surface of the pipe is breaking off in layers, evidenced by pitting and depressions in the pipe walls. This item is measured in linear metres or linear feet of pipe affected.

SPM—Parts missing (ea) Parts of a mechanical device are missing.

SSR—Scheduled service (ea) Maintenance services have been scheduled to repair a specific problem.

TA1, 2, 3, or 4—Tap abandoned 12–3, 3–6, 6–9, or 9–12 (ea) A tap, house connection (service lateral), or side connection is no longer in service; that is, no flow is currently entering the main line (pipe being inspected) from this connection, and it is not expected to be an active line again. The codes 12–3, 3–6, 6–9, and 9–12 refer, in clock positions, to the location of the tap along the pipe perimeter. For example, 12–3 is the 12 o'clock to 3 o'clock, or the top right (northeast) quadrant, of the pipe cross section.

TC1, 2, 3, or 4—Tap crushed 12–3, 3–6, 6–9, or 9–12 (ea) A tap, house connection (service lateral), or side connection has collapsed at the point it enters the main line (pipe being inspected) and is inhibiting flow to the main line. The codes 12–3, 3–6, 6–9, and 9–12 refer, in clock positions, to the location of the tap along the pipe perimeter. For example, 12–3 is the 12 o'clock to 3 o'-clock, or the top right (northeast) quadrant, of the pipe cross section.

TI1, 2, 3, or 4—Tap infiltration 12–3, 3–6, 6–9, or 9–12 (ea) At the location of a tap, house connection (service lateral), or side connection, water is leaking into the mainline (pipe being inspected) from an external source. The codes 12–3, 3–6, 6–9, and 9–12 refer, in clock positions, to the location of the tap along the pipe perimeter. For example, 12–3 is the 12 o'clock to 3 o'clock, or the top right (northeast) quadrant, of the pipe cross section.

TN1, 2, 3, or 4—Tap intruding 12–3, 3–6, 6–9, or 9–12 (ea) A tap, house connection (service lateral), or side connection that extends into the main line (pipe being inspected) is protruding beyond the interior face of the main line. This condition inhibits flow in the main line and is a source of debris accumulation that causes maintenance problems. The codes 12–3, 3–6, 6–9, and 9–12 refer, in clock positions, to the location of the tap along the pipe perimeter. For example, 12–3 is the 12 o'clock to 3 o'clock, or the top right (northeast) quadrant, of the pipe cross section.

TP1, 2, 3, or 4—Tap normal 12–3, 3–6, 6–9, or 9–12 (ea) An active tap, house connection (service lateral), or side connection entering the main line (pipe being inspected) has no known deficiencies. The codes 12–3, 3–6, 6–9, and 9–12 refer, in clock positions, to the location of the tap along the pipe perimeter. For example, 12–3 is the 12 o'clock to 3 o'clock, or the top right (northeast) quadrant, of the pipe cross section.

TR1, 2, 3, or 4—Tap with roots 12–3, 3–6, 6–9, or 9–12 (ea) Tree roots are entering the main line (pipe being inspected) at the location of a tap, house connection (service lateral), or side connection. The codes 12–3, 3–6, 6–9, and 9–12 refer, in clock positions, to the location of the tap along the pipe perimeter. For example, 12–3 is the 12 o'clock to 3 o'clock, or the top right (northeast) quadrant, of the pipe cross section.

TRP—Trench to be paved (m^2 or sq ft) A trench excavation to access a sewer pipe has been backfilled and requires paving. This item is measured in square metres or square feet of surface area that need paving.

TS1, 2, 3, or 4—Tap separated 12–3, 3–6, 6–9, or 9–12 (ea) A tap, house connection (service lateral), or side connection has moved differentially from the main line (pipe being inspected), causing a gap between the two. This separation is a potential source of infiltration and inflow, root intrusion, and structural deterioration of the pipes. The codes 12–3, 3–6, 6–9, and 9–12 refer, in clock positions, to the location of the tap along the pipe perimeter. For example, 12–3 is the 12 o'clock to 3 o'clock, or the top right (northeast) quadrant, of the pipe cross section.

TU1, 2, 3, or 4—Tap unknown 12–3, 3–6, 6–9, or 9–12 (ea) A pipe connecting to the main line (pipe being inspected) has an unknown source and requires further investigation. The codes 12–3, 3–6, 6–9, and 9–12 refer, in clock positions, to the location of the tap along the pipe perimeter. For example, 12–3 is the 12 o'clock to 3 o'clock, or the top right (northeast) quadrant, of the pipe cross section.

UNK—Unknown situation (ea) A problem in the inspection area is occurring but the source is unknown. Further investigation is required.

VOL—Void outside line (m^3 or cu yd) An empty space has formed adjacent to the pipe behind the wall that is visible with a television camera or manual inspection. If the size of this void can be assessed, it is measured in terms of volume or surface area. In many cases, however, the size of the void cannot be quantified.

VRM—Vermin, general (ea) Insects, bugs, or small animals not included in other classifications are present in the sewer system. This item is measured per occurrence.

WMB—Water main broken (ea) A water pipe has burst in the inspection area, evidenced by large quantities of water coming to the surface. Appropriate authorities are notified and the affected section of the water system is shut off for repairs.

ACTION/RESPONSE CODES

Code	Description	Unit
AST	Assist a crew	ea
BFV	Bait for vermin	ea
BKF	Backfill, tamp/jet	m^3 (cu yd)
BYP	Bypass: plug and pump	ea
CLB	Clean with ball	m (ft)
CLC	Clean with clam bucket	m^3 (cu yd)
CLD	Clean with dredge/drag	cu yd
CLG	Clean grate(s)	ea
CLJ	Clean with hydrojet	m (ft)
CLK	Clean with kite	m (ft)
CLS	Clean with scooter	m (ft)
CLV	Clean with vacuum	ea
CWR	Contract work required	ea

DCR	Dust for roaches	m^2 (sq ft)
DYE	Dye test	ea
EXC	Excavation	m^3 (cu yd)
FLU	Flush, no pressure	m (ft)
GRT	Grout seal	ea
GSK	Gasket seal/reseal	ea
INA	Inspect above ground	ea
INB	Inspect below ground	ea
INC	Inspection complete	ea
IST	*In situ* form	m (ft)
LAP	Line air test	ea
LNI	Line, new install	m (ft)
LOF	Locate overflow	ea
LRP	Line repair	m (ft)
LRR	Line remove and replace	m (ft)
LSL	Line, slip-line	m (ft)
MHF	Manhole, find/locate	ea
MHH	Manhole, height adjustment	ea
MHI	Manhole, installation	ea
MHR	Manhole, rechannel	ea
MOS	Mosquito control	ea
MRK	Mark for dig-up	ea
OTA	Other action(s)	ea
OWN	Owner notified	ea
PAD	Quiet cover/grate	ea
PAV	Pave/repave	sq ft
PLS	Plaster/replaster	sq ft
RCF	Repair cover/frame	ea
RCL	Replace cover/lid	ea
RDH	Rod by hand	m (ft)
RDP	Rod with power equipment	m (ft)
RDS	Rod with snake	m (ft)
RTC	Root cutting	m (ft)
RTF	Root foaming, chemical	m (ft)
RTS	Root soaking, chemical	m (ft)
SLA	Service lateral air test	ea
SLC	Service lateral connect	ea
SLS	Service lateral tap, seal	ea
SLT	Service lateral tap, new	ea
SMO	Smoke test	m (ft)
STI	Structure, new install	ea
STR	Structure, repair	ea
TVI	Television inspect	m (ft)
TVS	String television line	m (ft)

UBH	Unblock with hydrojet	ea
UBP	Unblock with plunger	ea
UBR	Unblock with rodding	ea
UBS	Unblock with snake	ea

Index

M

N

O

P

R